Mika Sato-Ilic, Lakhmi C. Jain

Innovations in Fuzzy Clustering

Studies in Fuzziness and Soft Computing, Volume 205

Editor-in-chief
Prof. Janusz Kacprzyk
Systems Research Institute
Polish Academy of Sciences
ul. Newelska 6
01-447 Warsaw
Poland
E-mail: kacprzyk@ibspan.waw.pl

Further volumes of this series can be found on our homepage: springer.com

Vol. 189. Hans Bandemer
Mathematics of Uncertainty, 2006
ISBN 3-540-28457-5

Vol. 190. Ying-ping Chen
Extending the Scalability of Linkage Learning Genetic Algorithms, 2006
ISBN 3-540-28459-1

Vol. 191. Martin V. Butz
Rule-Based Evolutionary Online Learning Systems, 2006
ISBN 3-540-25379-3

Vol. 192. Jose A. Lozano, Pedro Larrañaga, Iñaki Inza, Endika Bengoetxea (Eds.)
Towards a New Evolutionary Computation, 2006
ISBN 3-540-29006-0

Vol. 193. Ingo Glöckner
Fuzzy Quantifiers: A Computational Theory, 2006
ISBN 3-540-29634-4

Vol. 194. Dawn E. Holmes, Lakhmi C. Jain (Eds.)
Innovations in Machine Learning, 2006
ISBN 3-540-30609-9

Vol. 195. Zongmin Ma
Fuzzy Database Modeling of Imprecise and Uncertain Engineering Information, 2006
ISBN 3-540-30675-7

Vol. 196. James J. Buckley
Fuzzy Probability and Statistics, 2006
ISBN 3-540-30841-5

Vol. 197. Enrique Herrera-Viedma, Gabriella Pasi, Fabio Crestani (Eds.)
Soft Computing in Web Information Retrieval, 2006
ISBN 3-540-31588-8

Vol. 198. Hung T. Nguyen, Berlin Wu
Fundamentals of Statistics with Fuzzy Data, 2006
ISBN 3-540-31695-7

Vol. 199. Zhong Li
Fuzzy Chaotic Systems, 2006
ISBN 3-540-33220-0

Vol. 200. Kai Michels, Frank Klawonn, Rudolf Kruse, Andreas Nürnberger
Fuzzy Control, 2006
ISBN 3-540-31765-1

Vol. 201. Cengiz Kahraman (Ed.)
Fuzzy Applications in Industrial Engineering, 2006
ISBN 3-540-33516-1

Vol. 202. Patrick Doherty, Witold Łukaszewicz, Andrzej Skowron, Andrzej Szałas
Knowledge Representation Techniques: A Rough Set Approach, 2006
ISBN 3-540-33518-8

Vol. 203. Gloria Bordogna, Giuseppe Psaila (Eds.)
Flexible Databases Supporting Imprecision and Uncertainty, 2006
ISBN 3-540-33288-X

Vol. 204. Zongmin Ma (Ed.)
Soft Computing in Ontologies and Semantic Web, 2006
ISBN 3-540-33472-6

Vol. 205. Mika Sato-Ilic, Lakhmi C. Jain
Innovations in Fuzzy Clustering, 2006
ISBN 3-540-34356-3

Mika Sato-Ilic
Lakhmi C. Jain

Innovations in Fuzzy Clustering

Theory and Applications

Professor Dr. Mika Sato-Ilic
Department of Risk Engineering
Faculty of Systems and
Information Engineering
University of Tsukuba
Tennodai 1-1-1
Tsukuba, Ibaraki 305-8573
Japan
E-mail: mika@risk.tsukuba.ac.jp

Professor Dr. Lakhmi C. Jain
School of Electrical and
Information Engineering
University of South Australia
Adelaide
Mawson Lakes Campus
South Australia SA 5095
Australia
E-mail: Lakhmi.jain@unisa.edu.au

ISSN print edition: 1434-9922
ISSN electronic edition: 1860-0808
ISBN 978-3-642-07072-3 e-ISBN 978-3-540-34357-8

Springer is a part of Springer Science+Business Media
springer.com

Printed in The Netherlands

Cover design: Erich Kirchner, Heidelberg

Foreword

Clustering has been around for many decades and located itself in a unique position as a fundamental conceptual and algorithmic landmark of data analysis. Almost since the very inception of fuzzy sets, the role and potential of these information granules in revealing and describing structure in data was fully acknowledged and appreciated. As a matter of fact, with the rapid growth of volumes of digital information, the role of clustering becomes even more visible and critical. Furthermore given the anticipated human centricity of the majority of artifacts of digital era and a continuous buildup of mountains of data, one becomes fully cognizant of the growing role and an enormous potential of fuzzy sets and granular computing in the design of intelligent systems. In the recent years clustering has undergone a substantial metamorphosis. From being an exclusively *data* driven pursuit, it has transformed itself into a vehicle whose data centricity has been substantially augmented by the incorporation of domain knowledge thus giving rise to the next generation of *knowledge*-oriented and *collaborative* clustering.

Interestingly enough, fuzzy clustering exhibits a dominant role in many developments of the technology of fuzzy sets including fuzzy modeling, fuzzy control, data mining, pattern recognition, and image processing. When browsing through numerous papers on fuzzy modeling we can witness an important trend of a substantial reliance on fuzzy clustering being regarded as the general development tool. The same central position of fuzzy clustering becomes visible in pattern classifiers and neurofuzzy systems. All in all, it becomes evident that further progress in fuzzy clustering is of vital relevance and benefit to the overall progress of the area of fuzzy sets and their applications.

In this context, the book by Professors Mika Sato-Ilic and Lakhmi Jain is an authoritative and timely treatise on the important subject of fuzzy clustering. The authors, who are experts in the field, put the subject matter into an innovative and attractive framework and provide the readers with a unique and cohesive view of the dynamically expanding discipline and its applications. The treatment of the topic covered in the book is appealing in many different ways. On one hand, the expertise of the authors in the discipline of fuzzy clustering shows up very profoundly and as a result of that the reader could fully enjoy the didactic aspects of the material. On the other hand, the coverage of the original ideas Professors Sato-Ilic and Jain deserve a full credit for, comes hand in hand with a systematic exposure whose enhancements realized within the setting of fuzzy sets become truly beneficial. The authors embark on important and timely facets of fuzzy clustering involving principal component analysis, regression models and kernel-based and self-organizing fuzzy clustering. They also include a thorough and constructive discussion on the evaluation of results of fuzzy clustering – a subject that is always highly welcome.

Unquestionably, this book offers a number of new concepts which, in my opinion, will promote further developments in the area of intelligent data analysis and contribute to the continuous growth and enrichment of the technology of fuzzy sets.

Once again, Springer's series "Studies in Fuzziness and Soft Computing" gives readers yet another innovative and authoritative volume for which there is great need. Authored by renowned experts, this book serves as a testimony to the fertile conceptual and algorithmic wealth of fuzzy sets and their leading and highly stimulating role in human-centric endeavours.

Edmonton, Alberta, Canada
April, 2006

Professor Witold Pedrycz
President
International Fuzzy System Association (IFSA)

Preface

There is a great interest in clustering techniques due to the vast amount of data generated in every field including business, health sciences, engineering and aerospace. It is essential to extract useful information from the data. Clustering techniques are widely used in pattern recognition and related applications. This research monograph presents the most recent advances in clustering techniques and their applications.

Conventionally, fuzzy multivariate analysis has been proposed together with the issue of positively introducing uncertainty in real data and the methodology. Fuzzy clustering is one method which can capture the uncertainty situation of real data and it is well known that fuzzy clustering can obtain a robust result as compared with conventional hard clustering. Following the emphasis on the general problem of data analysis, which is a solution able to analyze a huge amount of complex data, the merit of fuzzy clustering is then presented. We describe fuzzy clustering methods, which are methods of fuzzy multivariate analysis, along with several hybrid methods of fuzzy clustering and conventional multivariate analysis recently proposed by us. These are based on the idea that the multiple merits of several methods can cope with the inherent classification structures.

Fuzzy multivariate analysis and conventional multivariate analysis has the common purpose of analyzing huge and complex data. The former emphasizes the aspect of practical use and the latter emphasizes the theory. Moreover, various data analyses have been proposed in the area of soft computing which includes fuzzy logic. The methods of data analysis using soft computing include fuzzy multivariate analysis, neural networks, and support vector machine. These have a similar methodology of data analyses discussed in conventional statistical science. On the other hand, in statistical data analysis, methodologies especially related to these areas, include symbolic data analysis, functional data analysis, and several kinds of non-linear multivariate analyses.

In these methodologies there is a difference, that is dependant on whether they use a fuzzy membership function or a density function. This choice is related with the difference of definitions of uncertainty, that is, the difference between fuzzy structure and probabilistic structure. In either case the purpose of the analyses is to estimate the non-linear parameters which define the properties of the uncertainty.

There is research which aims to fuse data analyses in soft computing arising from the aspect of practical use and data analyses from the view point of statistical science, that is, from the aspect of theory. This type of research aims to investigate the idea that both follow different courses, but both arrive at the same point. These are not discussions on the difference between "probability" and "fuzziness" which has been conventionally discussed. They explore the mathematical differences and enable us to construct methods of data analysis based on theory and practice.

The first chapter introduces the basic concepts and techniques considered in this book. We introduce fuzzy set and fuzzy clustering in chapter 1. As a part of the methodology, we introduce principal component analysis based on fuzzy clustering and regression analysis based on fuzzy clustering. These hybrid techniques of fuzzy data analysis and conventional multivariate analysis are presented in chapters 2 and 3. The second chapter presents fuzzy clustering based principal component analysis. This chapter describes weighted principal component analysis using fuzzy clustering. The third chapter presents fuzzy clustering based regression analysis. It includes novel hybrid techniques of fuzzy clustering and regression methods. In order to obtain more accurate results, we introduce attempts to use a kernel method for data analysis. These methods are described in chapter 4. The fourth chapter presents kernel based fuzzy clustering. It describes the extended fuzzy cluster loadings using the kernel method in the higher dimension space. The superiority of the kernel based fuzzy cluster loading is demonstrated using practical examples. In chapter 5, we discuss the evaluation of fuzzy clustering methods related to the validity of clustering results. We present evaluation techniques of results of fuzzy clustering using homogeneity analysis. Fuzzy clustering provides natural groups within the given observations using an assumption of a fuzzy subset defined fuzzy clusters. Each object can belong to several clusters with several degrees and boundaries of clusters. Thus, it becomes difficult to interpret the clusters and evaluation techniques play an important role in fuzzy clustering. In order to obtain defuzzified results, chapter 6 discusses a self-organized fuzzy clustering, and a hybrid method of fuzzy clustering and multidimensional scaling based on a self-organized dissimilarity or a self-organized similarity between a pair of objects. Examples are given in the end of the chapter to demonstrate the validity of our approach.

These methods aim to capture more accurate results and to avoid noise of data, by using the essential property of fuzzy clustering which is obtaining the degree of belongingness of objects to clusters as continuous values and by applying the classification structure to other methods.

Arising from the similarity between fuzzy structure and probabilistic structure, many similar methods have been proposed both in the areas of fuzzy logic and statistic theory. The methods introduced in this book are a small part and consequently it is important to inspect in detail these relationships in the future.

We are indebted to the researchers in the area of clustering techniques for their numerous publications. We are benefited by their work. We wish to express our gratitude to Dr Neil Allen for his critical reading of the manuscript. The editorial assistance by the publisher is also acknowledged.

Adelaide
May 2005

Mika Sato-Ilic
Lakhmi C. Jain

Contents

1. Introduction to Fuzzy Clustering

1.1 Fuzzy Logic

Fuzzy sets [68] were first proposed as a method of representing the uncertainty inherent in real data. This is an extension of conventional set theory. However, in the case of sets treated by conventional set theory, the elements in a set have to be judged as to whether the elements belong to the set or not. In the case of fuzzy sets, whether the elements belong to the set or not is unclear. In order to represent this mathematically, using the degree of the belongingness of each element to the set, fuzzy subsets are defined as follows:

Definition 1. (Fuzzy Subset): A fuzzy subset A is a set characterized by the following membership function:

$$\mu_A \, : \, X \to [0,1].$$

For any elements $\boldsymbol{x} \in X$, $\mu_A(\boldsymbol{x}) \in [0,1]$ shows the degree of belongingness (grade) of $\boldsymbol{x}$ for a fuzzy set A. If the grade $\mu_A(\boldsymbol{x})$ is close to 1, then it means a situation of a greater degree of belongingness of the element $\boldsymbol{x}$ to the set A. If the degree is close to 0, then the degree of belongingness of $\boldsymbol{x}$ to A becomes small.

In particular, if $\mu_A(\boldsymbol{x})$ is only 0 or 1 for any $\boldsymbol{x}$, that is, $\mu_A(\boldsymbol{x}) \in \{0,1\}$, then this membership function is a characteristic function in usual set theory and A is a usual subset. In this sense, a usual set is a special case of a fuzzy set and the fuzzy set is an extension of the usual set.

1.2 Fuzzy Clustering

Conventional clustering means classifying the given observation as exclusive subsets (clusters). That is, we can discriminate clearly whether

Mika Sato-Ilic and Lakhmi C. Jain: *Innovations in Fuzzy Clustering*, StudFuzz **205**, 1–8 (2006)
www.springerlink.com

an object belongs to a cluster or not. However, such a partition is insufficient to represent many real situations. Therefore, a fuzzy clustering method is offered to construct clusters with uncertain boundaries, so this method allows that one object belongs to some overlapping clusters to some degree. In other words, the essence of fuzzy clustering is to consider not only the belonging status to the clusters, but also to consider to what degree do the objects belong to the clusters. There is merit in representing the complex data situations of real data.

Suppose

$$X = \{\boldsymbol{x}_1, \boldsymbol{x}_2, \ldots, \boldsymbol{x}_n\}$$

is a given set of n objects and K $(K = 1, \ldots, n; K \in N)$ is a number of clusters. Where, N is a set of all natural numbers.

Then a fuzzy cluster which is a fuzzy subset in X is defined as follows:

$$\mu_k : X \to [0,1], \quad k = 1, \ldots, K.$$

The fuzzy grade for each fuzzy cluster k is denoted as:

$$u_{ik} \equiv \mu_k(\boldsymbol{x}_i), \; i = 1, \ldots, n, \quad k = 1, \ldots, K.$$

That is, u_{ik} shows the degree of belongingness of an object i to a cluster k. In general, u_{ik} satisfies the following conditions:

$$u_{ik} \in [0,1], \; \forall i, k; \; \sum_{k=1}^{K} u_{ik} = 1, \; \forall i. \tag{1.1}$$

The state of fuzzy clustering is represented by a partition matrix $U = (u_{ik})$ and a set of the matrixes is shown as:

$$M_{fnK} = \{U \in R^{nK} | \; u_{ik} \text{ satisfies (1.1) } \forall i, k\}. \tag{1.2}$$

In particular, if for any i and k, $u_{ik} \in \{0, 1\}$, that is a case of hard clustering, the set of the partition matrixes U is

$$M_{nK} = \{U \in M_{fnK} | u_{ik} \in \{0,1\}, \forall i, k\},$$

and we obtain the following relationship

$$M_{nK} \subset M_{fnK}.$$

That is, hard clustering is a special case of fuzzy clustering.

1.2.1 Fuzzy C-Means (FCM) and Relational Fuzzy C-Means (RFCM)

Fuzzy c-means (FCM) [4] is one of the methods of fuzzy clustering. FCM is the method which minimizes the weighted within-class sum of squares:

$$J(U, \boldsymbol{v}_1, \ldots, \boldsymbol{v}_K) = \sum_{i=1}^{n} \sum_{k=1}^{K} (u_{ik})^m \, d^2(\boldsymbol{x}_i, \boldsymbol{v}_k), \tag{1.3}$$

where $\boldsymbol{v}_k = (v_{ka}),\ k = 1, \ldots, K,\ a = 1, \ldots, p$ denotes the values of the centroid of a cluster k, $\boldsymbol{x}_i = (x_{ia}),\ i = 1, \ldots, n,\ a = 1, \ldots, p$ is i-th object with respect to p variables, and $d^2(\boldsymbol{x}_i, \boldsymbol{v}_k)$ is the square Euclidean distance between $\boldsymbol{x}_i$ and $\boldsymbol{v}_k$. The exponent m which determines the degree of fuzziness of the clustering is chosen from $(1, \infty)$ in advance. The purpose of this is to obtain the solutions U and $\boldsymbol{v}_1, \ldots, \boldsymbol{v}_K$ which minimize equation (1.3). The solutions are obtained by Picard iteration of the following expressions:

$$u_{ik} = \frac{1}{\sum_{l=1}^{K} \{d(\boldsymbol{x}_i, \boldsymbol{v}_k)/d(\boldsymbol{x}_i, \boldsymbol{v}_l)\}^{\frac{2}{m-1}}}, \quad \boldsymbol{v}_k = \frac{\sum_{i=1}^{n} (u_{ik})^m \boldsymbol{x}_i}{\sum_{i=1}^{n} (u_{ik})^m},$$

$$i = 1, \ldots, n, \quad k = 1, \ldots, K.$$

From equation (1.3), we can rewrite this as:

$$\begin{aligned} J(U, \boldsymbol{v}_1, \ldots, \boldsymbol{v}_K) &= \sum_{i=1}^{n} \sum_{k=1}^{K} (u_{ik})^m \, d^2(\boldsymbol{x}_i, \boldsymbol{v}_k) \\ &= \sum_{i=1}^{n} \sum_{k=1}^{K} (u_{ik})^m \, (\boldsymbol{x}_i - \boldsymbol{v}_k, \boldsymbol{x}_i - \boldsymbol{v}_k) \\ &= \sum_{i=1}^{n} \sum_{k=1}^{K} (u_{ik})^m \, (\boldsymbol{x}_i - \boldsymbol{h}_k + \boldsymbol{h}_k - \boldsymbol{v}_k, \\ &\qquad \boldsymbol{x}_i - \boldsymbol{h}_k + \boldsymbol{h}_k - \boldsymbol{v}_k) \\ &= \sum_{i=1}^{n} \sum_{k=1}^{K} (u_{ik})^m \, [(\boldsymbol{x}_i - \boldsymbol{h}_k, \boldsymbol{x}_i - \boldsymbol{h}_k) \\ &\quad + 2(\boldsymbol{x}_i - \boldsymbol{h}_k, \boldsymbol{h}_k - \boldsymbol{v}_k) + (\boldsymbol{h}_k - \boldsymbol{v}_k, \boldsymbol{h}_k - \boldsymbol{v}_k)], \end{aligned}$$

where $(\cdot, \cdot)$ denotes a real inner product.

If we assume

$$\boldsymbol{h}_k = \frac{\sum_{i=1}^{n} (u_{ik})^m \boldsymbol{x}_i}{\sum_{i=1}^{n} (u_{ik})^m},$$

then the minimizer of equation (1.3) is shown as:

$$J(U) = \sum_{k=1}^{K} \left(\sum_{i=1}^{n} \sum_{j=1}^{n} ((u_{ik})^m (u_{jk})^m d_{ij}) / (2 \sum_{s=1}^{n} (u_{sk})^m) \right), \tag{1.4}$$

using

$$\begin{aligned} & 2 \sum_{s=1}^{n} (u_{sk})^m \sum_{i=1}^{n} (u_{ik})^m (\boldsymbol{x}_i - \boldsymbol{h}_k, \boldsymbol{x}_i - \boldsymbol{h}_k) \\ & = \sum_{i=1}^{n} \sum_{j=1}^{n} (u_{ik})^m (u_{jk})^m (\boldsymbol{x}_i - \boldsymbol{x}_j, \boldsymbol{x}_i - \boldsymbol{x}_j), \end{aligned}$$

and $d_{ij} = d(\boldsymbol{x}_i, \boldsymbol{x}_j)$. The equation (1.4) is the objective function of the relational fuzzy c-means [22]. When $m = 2$, equation (1.4) is the objective function of the FANNY algorithm. The details of the FANNY algorithm are given in [27].

1.2.2 RFCM for 3-Way Data

In this section, we describe two RFCM methods for 3-way data [41]. The 3-way data, which is observed by the values of dissimilarity with respect to n objects for T times, are denoted by the following:

$$\Delta^{(t)} = (\delta_{ij}^{(t)}), \quad i, j = 1, \ldots, n, \quad t = 1, \ldots, T.$$

The purpose of this clustering is to classify n objects into K fuzzy clusters $C_1, \ldots, C_K$ which are the fuzzy subsets of the set of objects $O = \{o_1, \ldots, o_n\}$. From equation (1.4), the goodness of clustering in the t-th time is given by a modified equation of the sum of extended within-class dispersion,

$$F^{(t)}(U) = \sum_{k=1}^{K} \left(\sum_{i=1}^{n} \sum_{j=1}^{n} ((u_{ik})^m (u_{jk})^m \delta_{ij}^{(t)}) / (2 \sum_{s=1}^{n} (u_{sk})^m) \right). \tag{1.5}$$

If there exists a solution U which minimizes all $F^{(t)}$ $(t = 1, \ldots, T)$, then it is the best or dominant solution. Usually such a solution does not exist. Therefore this problem, a clustering for 3-way data, becomes a multicriteria optimization problem. We now assume that Φ is a set of feasible solutions U. We define a single clustering criterion by the weighted sum of $F^{(t)}$, that is, for $w^{(t)} > 0$,

$$F(U) = \sum_{t=1}^{T} w^{(t)} F^{(t)}(U). \tag{1.6}$$

By theorem 1 and theorem 2, the equation (1.6) shows that the problem of finding a Pareto efficient solution of $(\Phi, F^{(1)}, \ldots, F^{(T)})$ is reduced to a usual nonlinear optimization problem.

Theorem 1.[9]: Let $\boldsymbol{x}^* \in \Phi$ be an optimum solution of the weighting method with the weights $w^{(t)} > 0$, $(t = 1, \ldots, T)$. Then $\boldsymbol{x}^*$ is a Pareto efficient solution of the multicriteria problem $(\Phi, F^{(1)}, \ldots, F^{(T)})$.

Theorem 2.[9]: Let $F^{(t)}(\boldsymbol{x})$, $(t = 1, \ldots, T)$ be convex functions. If $\boldsymbol{x}^* \in \Phi$ is a Pareto efficient solution of the multicriteria problem $(\Phi, F^{(1)}, \ldots, F^{(T)})$, then $\boldsymbol{x}^*$ is an optimum solution of the weighting method with the weights $w^{(t)} \geq 0$.

One should note that the fuzzy clusters or fuzzy subsets, are uniquely determined through the times $t = 1, \ldots, T$. The purpose of the clustering (1.6) is to get common clusters through all the time points. That is, the grades u_{ik} are independent of the time points. If the different clusters are determined at every time point, such that the grades u_{ik} are dependent on the time point, this clustering is considered to be a usual clustering for 2-way data at each time point. Therefore, in this case, it is not necessary to consider the multicriteria problem.

In order to calculate the matrix of the degree of belongingness of objects to clusters, u_{ik}, shown in equation (1.1), we use the following algorithm:

(Step 1) Fix K, $(2 \leq K < n)$, $w^{(t)}$, $(t = 1, \ldots, T)$, and m, $(m \in (1, \infty))$.

(Step 2) Initialize $U(0) = (u_{ik}(0))$, $(i = 1, \ldots, n; k = 1, \ldots, K)$. And set the step number $q = 0$.

(Step 3) Set $q = q + 1$. Calculate the c-means vectors

$$\boldsymbol{v}_k(q-1) = \frac{((u_{1k}(q-1))^m, (u_{2k}(q-1))^m, \ldots, (u_{nk}(q-1))^m)'}{\sum_{i=1}^{n}(u_{ik}(q-1))^m},$$

$$k = 1, \ldots, K$$

(Step 4) Calculate the weighted dissimilarities among objects

$$\Delta(q-1) = (\sum_{t=1}^{T} w^{(t)} \delta_{ij}^{(t)}(q-1)) \equiv (\delta_{ij}(q-1)),\ i, j = 1, \ldots, n,$$

and calculate

$$\delta_{ik}(q-1) = (\Delta(q-1)\boldsymbol{v}_k(q-1))_i - (\boldsymbol{v}_k'(q-1)\Delta(q-1)\boldsymbol{v}_k(q-1))/2,$$

where $(\cdot)_i$ shows i-th row of $(\cdot)$.

(Step 5) Calculate the degree of belongingness of the i-th object to the k-th cluster,

$$u_{ik}(q-1) = \frac{1}{\sum_{l=1}^{K}(\delta_{ik}(q-1)/\delta_{il}(q-1))^{\frac{2}{m-1}}}.$$

(Step 6) If $\|U(q) - U(q-1)\| \leq \varepsilon$, then stop, or otherwise go to Step 3. Where, ε is a threshold.

The above algorithm is essentially the same as the algorithm of RFCM without the calculation of dissimilarity matrix at step 4.

Next, we describe another relational fuzzy c-means for 3-way data named Dynamic Relational Fuzzy C-Means (DRFCM). The data is observed by the values of dissimilarity with respect to n objects for T times, and the dissimilarity matrix of t-th time is shown by $\Delta^{(t)} = (\delta_{ij}^{(t)})$. Then a $Tn \times Tn$ matrix $\tilde{\Delta}$ is denoted as follows:

$$\tilde{\Delta} = \begin{bmatrix} \Delta^{(1)} & \Delta^{(12)} & \ldots & \Delta^{(1T)} \\ \Delta^{(21)} & \Delta^{(2)} & \ldots & \Delta^{(2T)} \\ \vdots & \vdots & \ddots & \vdots \\ \Delta^{(T1)} & \Delta^{(T2)} & \ldots & \Delta^{(T)} \end{bmatrix}, \tag{1.7}$$

where the diagonal matrix is the $n \times n$ matrix $\Delta^{(t)}$. $\Delta^{(rt)}$ is an $n \times n$ matrix and the element is defined as

$$\delta_{ij}^{(rt)} \equiv m(\delta_{ij}^{(r)}, \delta_{ij}^{(t)}),$$

where $\delta_{ij}^{(t)}$ is the (i,j)-th element of the matrix $\Delta^{(t)}$. $m(x,y)$ is an average function from the product space $[0,1] \times [0,1]$ to $[0,1]$ and satisfies the following conditions:

(1) $\min\{x,y\} \leq m(x,y) \leq \max\{x,y\}$

(2) $m(x,y) = m(y,x)$:(symmetry)

(3) $m(x,y)$ is increasing and continuous

Moreover, from (1), $m(x,y)$ satisfies the following condition:

(4) $m(x,x) = x$:(idempotency)

The examples of the average function are shown in table 1.1.

Table 1.1 Examples of Average Function

Harmonic Mean	$\frac{2xy}{x+y}$
Geometric Mean	$\sqrt{xy}$
Arithmetic Mean	$\frac{x+y}{2}$
Dual of Geometric Mean	$1-\sqrt{(1-x)(1-y)}$
Dual of Harmonic Mean	$\frac{x+y-2xy}{2-x-y}$

For the element $\tilde{\delta}_{ij}$ of $\tilde{\Delta}$, a new fuzzy clustering criterion which is named the Dynamic Relational Fuzzy C-Means method is as follows:

$$F(\tilde{U}) = \sum_{k=1}^{K} \left(\sum_{i=1}^{Tn} \sum_{j=1}^{Tn} ((\tilde{u}_{ik})^m (\tilde{u}_{jk})^m \tilde{\delta}_{ij}) / (2 \sum_{s=1}^{Tn} (\tilde{u}_{sk})^m) \right), \quad (1.8)$$

where $i, j, s = 1, \ldots, Tn$, and K is a number of clusters. If $i^{(t)} \equiv i - n(t-1)$, $n(t-1)+1 \le i \le tn$, $t = 1, \ldots, T$, then $i^{(t)} = 1, \ldots, n$, $\forall t$ and $i^{(t)}$ shows the object number. That is, $\tilde{u}_{i^{(t)}k}$ shows a degree of belongingness of an object i to a cluster k for time t. So, we can obtain the dynamic results through the times on the same clusters and then the results will be comparable.

The degree of belongingness of objects to clusters over times, $\tilde{u}_{ik}$ in equation (1.8), can be obtained using the following algorithm:

(Step 1) Fix K, $(2 \le K < n)$, and m, $(m \in (1, \infty))$.
(Step 2) Initialize $\tilde{U}(0) = (\tilde{u}_{ik}(0))$, $(i = 1, \ldots, Tn; k = 1, \ldots, K)$. And set the step number $q = 0$.
(Step 3) Set $q = q + 1$. Calculate the c-means vectors

$$\tilde{\boldsymbol{v}}_k(q-1) = \frac{((\tilde{u}_{1k}(q-1))^m, (\tilde{u}_{2k}(q-1))^m, \ldots, (\tilde{u}_{Tn,k}(q-1))^m)'}{\sum_{i=1}^{Tn} (\tilde{u}_{ik}(q-1))^m},$$

$$k = 1, \ldots, K$$

(Step 4) Calculate the dissimilarities among objects

$$\tilde{\Delta}(q-1) = (\tilde{\delta}_{ij}(q-1)), \ i, j = 1, \ldots, Tn,$$

and calculate

$$\tilde{\delta}_{ik}(q-1) = (\tilde{\Delta}(q-1)\tilde{\boldsymbol{v}}_k(q-1))_i - (\tilde{\boldsymbol{v}}_k'(q-1)\tilde{\Delta}(q-1)\tilde{\boldsymbol{v}}_k(q-1))/2,$$

where $(\cdot)_i$ shows i-th row of $(\cdot)$.
(Step 5) Calculate the degree of belongingness of the i-th object to the k-th cluster,

$$\tilde{u}_{ik}(q-1) = \frac{1}{\sum_{l=1}^{K} (\tilde{\delta}_{ik}(q-1)/\tilde{\delta}_{il}(q-1))^{\frac{2}{m-1}}}.$$

(Step 6) If $\|\tilde{U}(q) - \tilde{U}(q-1)\| \le \varepsilon$, then stop, or otherwise go to Step 3. Where, ε is a threshold.

2. Fuzzy Clustering based Principal Component Analysis

In this chapter, we describe Weighted Principal Component Analyses (WPCA) [48], [52], [56] using the result of fuzzy clustering [5], [36]. The Principal Component Analysis (PCA) [1], [26] is a widely used and well-known data analysis technique. However problems arise, when the data does not have a structure that PCA can capture and thus we cannot obtain any satisfactory results. For the most part, this is due to the uniformity of the data structure, which means we cannot find any significant proportion or accumulated proportion for the principal components obtained.

In order to solve this problem, we use the classification structure and degree of belongingness of objects to the clusters, which is obtained as the fuzzy clustering result. By introducing pre-classification and the degree of belongingness to the data, we can transform the data into clearer structured data and avoid the noise in the data.

2.1 Background of Weighted Principal Component Analysis

The PCA is widely used in multivariate analysis. The particular feature of PCA can represent the main tendency of observed data compactly. PCA therefore has been used as a method to follow up a clue when any significant structure in the data is not obvious. Beside the importance of PCA as an exploratory method, several extended methods of PCA have been proposed with the view that PCA can bring additional characteristics of the observed data into a form suitable for analysis. For example, constrained PCA (CPCA) [60], [61], nonlinear PCA [57], and the time dependent principal component analysis [3] are typical examples.

The methods described in this chapter are based on the idea that we fully utilize PCA's ability through the introduction of pre-information

Mika Sato-Ilic and Lakhmi C. Jain: *Innovations in Fuzzy Clustering*, StudFuzz **205**, 9–44 (2006)
www.springerlink.com

of the classification structure of observed data. In these methods, we introduce two data structures which are classification structure and principal component structure. One of them is used for weights and the other is used for self analysis. We can reduce the risk of a wrong assumption of the introduced data structure, comparing the conventional method which assumes only one data structure on the observation.

The pre-information of the data is represented as weights based on a fuzzy clustering result. The main difference between conventional PCA and WPCA based on fuzzy clustering is the introduction of degree as weights for clusters. The weights are represented by the product of the degree of belongingness of objects to fuzzy clusters when an object is fixed. Due to a property of the algebraic product and the condition of the degree of belongingness, these weights can show to what degree each object has a classification structure. We show that the estimates of principal components for WPCA based on fuzzy clustering are obtained in a similar way as in conventional PCA. The time dependent principal component analysis [3] also adds weights to PCA, but does not use the classification structure of the data. Moreover, a possibilistic regression model [62], the switching regression model [23], and the weighted regression analysis based on fuzzy clustering [47], [55] are among several examples of research that introduce the concept of fuzziness to the regression model. Several models using fuzziness for the regression model are described in the next chapter.

2.2 Principal Component Analysis

When we observe the objects with respect to variables, PCA assumes that there are several main components (factors) caused by the relationship of the variables. The purpose of PCA is to obtain the components from the data and capture the features of the data.

The observed data are values of n objects with respect to p variables denoted by the following:

$$X = (x_{ia}), \quad i = 1, \ldots, n, \; a = 1, \ldots, p. \tag{2.1}$$

Suppose that X is centered with respect to p variables and the first principal component $\boldsymbol{z}_1$ is defined as the following linear combination:

$$\boldsymbol{z}_1 = X\boldsymbol{l}_1, \tag{2.2}$$

where $\boldsymbol{l}_1 = \begin{pmatrix} l_{11} \\ \vdots \\ l_{1p} \end{pmatrix}$.

The purpose of PCA is to estimate $\boldsymbol{l}_1$ which maximizes the variance of $\boldsymbol{z}_1$ under the condition of $\boldsymbol{l}_1'\boldsymbol{l}_1 = 1$. $(\cdot)'$ means transposition of $\cdot$. The variance of $\boldsymbol{z}_1$ is as follows:

$$V\{\boldsymbol{z}_1\} = V\{X\boldsymbol{l}_1\} = \boldsymbol{l}_1'V\{X\}\boldsymbol{l}_1 = \boldsymbol{l}_1'\Sigma\boldsymbol{l}_1, \tag{2.3}$$

where, $V\{\cdot\}$ shows the variance of $\cdot$ and Σ is a variance-covariance matrix of X. Using Lagrange's method of indeterminate multiplier, the following condition is needed for $\boldsymbol{l}_1$ to have a non-trivial solution.

$$|\Sigma - \lambda I| = 0, \tag{2.4}$$

where λ is an indeterminate multiplier and I is a unit matrix. Since equation (2.4) is the characteristic equation, λ is obtained as an eigen-value of Σ. Using equation (2.4) and the condition $\boldsymbol{l}_1'\boldsymbol{l}_1 = 1$, we obtain:

$$\boldsymbol{l}_1'\Sigma\boldsymbol{l}_1 = \lambda\boldsymbol{l}_1'\boldsymbol{l}_1 = \lambda. \tag{2.5}$$

Using equations (2.3) and (2.5), we obtain $V\{\boldsymbol{z}_1\} = \lambda$. So, $\boldsymbol{l}_1$ is determined as the corresponding eigen-vector for the maximum eigen-value of Σ.

In order to define the second principal component $\boldsymbol{z}_2$, we use the following linear combination:

$$\boldsymbol{z}_2 = X\boldsymbol{l}_2, \tag{2.6}$$

where $\boldsymbol{l}_2 = \begin{pmatrix} l_{21} \\ \vdots \\ l_{2p} \end{pmatrix}$. We need to estimate $\boldsymbol{l}_2$ which maximizes the variance of $\boldsymbol{z}_2$ under the condition of $\boldsymbol{l}_2'\boldsymbol{l}_2 = 1$ and covariance of $\boldsymbol{z}_1$ and $\boldsymbol{z}_2$ is 0. That is, $\boldsymbol{z}_1$ and $\boldsymbol{z}_2$ are mutually uncorrelated. If we denote the covariance between $\boldsymbol{z}_1$ and $\boldsymbol{z}_2$ as $\mathrm{cov}\{\boldsymbol{z}_1, \boldsymbol{z}_2\}$, then from equation (2.5), this condition can be represented as follows:

$$\text{cov}\{\boldsymbol{z}_1, \boldsymbol{z}_2\} = \text{cov}\{X\boldsymbol{l}_1, X\boldsymbol{l}_2\} = \boldsymbol{l}_1'\text{cov}\{X, X\}\boldsymbol{l}_2 = \boldsymbol{l}_1'\Sigma\boldsymbol{l}_2 = \lambda\boldsymbol{l}_1'\boldsymbol{l}_2 = 0.$$

That is, we need to estimate $\boldsymbol{l}_2$ which satisfy $(\Sigma - \lambda I)\boldsymbol{l}_2 = 0$ under the conditions $\boldsymbol{l}_2'\boldsymbol{l}_2 = 1$ and $\boldsymbol{l}_1'\boldsymbol{l}_2 = 0$. From the fact that the eigen-vectors corresponding to the different eigen-values are mutually orthogonal, $\boldsymbol{l}_2$ is found as the eigen-vector corresponding to the second largest eigen-value of Σ. If we get s $(s \leq p)$ different eigen-values of Σ, according to the above, we find the s principal components.

An indicator which can show how many principal components are needed to explain the data satisfactory, the following has been proposed. The proportion of the α-th principal component P_α is defined as:

$$P_\alpha = \frac{\lambda_\alpha}{\text{tr}(\Sigma)}. \tag{2.7}$$

From $V\{\boldsymbol{z}_\alpha\} = \lambda_\alpha$ and $\sum_{\alpha=1}^{p} \lambda_\alpha = \text{tr}(\Sigma)$, P_α can indicate the importance of the α-th principal component. Also, the accumulated proportion from P_1 to P_s is defined as:

$$P = \sum_{\alpha=1}^{s} P_\alpha. \tag{2.8}$$

From equations (2.7) and (2.8), $0 < P \leq 1$. In order to obtain an interpretation of the obtained principal components, the factor loading $r_{\alpha,j}$ is used. It is defined as a correlation coefficient between the α-th principal component $\boldsymbol{z}_\alpha$ and the j-th variable $\boldsymbol{x}_j$ as follows:

$$r_{\alpha,j} = \frac{\text{cov}\{\boldsymbol{z}_\alpha, \boldsymbol{x}_j\}}{\sqrt{V\{\boldsymbol{z}_\alpha\}V\{\boldsymbol{x}_j\}}} = \frac{\sqrt{\lambda_\alpha}l_{\alpha j}}{\sqrt{\sigma_{jj}}}, \tag{2.9}$$

where σ_{jj} is variance of $\boldsymbol{x}_j$.

2.3 Weighted Principal Component Analysis (WPCA)

We apply a fuzzy clustering method to the data matrix X shown in equation (2.1) and obtain the degree of belongingness $U = (u_{ik})$, $i = 1, \ldots, n$, $k = 1, \ldots, K$ shown in equation (1.1). We show the estimate of $U = (u_{ik})$ obtained as $\hat{U} = (\hat{u}_{ik})$. Using $\hat{U}$, we define the weight matrix W as follows:

$$W = \begin{pmatrix} \prod_{k=1}^{K} \hat{u}_{1k}^{-1} & \cdots & 0 \\ \vdots & \ddots & \vdots \\ 0 & \cdots & \prod_{k=1}^{K} \hat{u}_{nk}^{-1} \end{pmatrix} \equiv \begin{pmatrix} w_1 & \cdots & 0 \\ \vdots & \ddots & \vdots \\ 0 & \cdots & w_n \end{pmatrix}. \tag{2.10}$$

Then we introduce the following weighted matrix WX:

$$WX = \begin{pmatrix} w_1 & \cdots & 0 \\ \vdots & \ddots & \vdots \\ 0 & \cdots & w_n \end{pmatrix} \begin{pmatrix} x_{11} & \cdots & x_{1p} \\ \vdots & \ddots & \vdots \\ x_{n1} & \cdots & x_{np} \end{pmatrix} = \begin{pmatrix} w_1 x_{11} & \cdots & w_1 x_{1p} \\ \vdots & \ddots & \vdots \\ w_n x_{n1} & \cdots & w_n x_{np} \end{pmatrix}. \tag{2.11}$$

In order to avoid 0^{-1}, we replace equation (1.1) with the following conditions:

$$u_{ik} \in (0, 1),\ \forall i, k;\ \sum_{k=1}^{K} u_{ik} = 1,\ \forall i. \tag{2.12}$$

From the property of the algebraic product and the conditions (2.12), it follows that if $\hat{u}_{ik} = \frac{1}{K}$, for $\exists i$, $\forall k$, then $\prod_{k=1}^{K} \hat{u}_{ik}^{-1} (= w_i)$, $(\exists i)$ takes minimum value. If $\hat{u}_{ik}$ is close to 1 for $\exists i$, $\exists k$, then $\prod_{k=1}^{K} \hat{u}_{ik}^{-1} (= w_i)$, $(\exists i)$ is close to maximum value. That is, the weight w_i shows that if the status of belonging of an object i to the clusters is clearer, the weight becomes larger. Otherwise, if the belonging status of an object i is unclear, that is, the classification structure of the object is close to uniformity, then the weight becomes small. Hence, WX shows that clearer objects under the classification structure have larger values and that the objects which are vaguely situated under the classification structure are avoided. That is, those which do not have any significant relation to the classification structure are treated as noise.

Suppose $\tilde{\boldsymbol{z}}_1$ is the first principal component of the transformed data $WX \equiv \tilde{X}$ shown in equation (2.11). We center $\tilde{X}$ with respect to p variables and replace it as $\tilde{X}$. Then $\tilde{\boldsymbol{z}}_1$ is defined as:

$$\tilde{\boldsymbol{z}}_1 = \tilde{X}\tilde{\boldsymbol{l}}_1, \tag{2.13}$$

where $\tilde{\boldsymbol{l}}_1 = \begin{pmatrix} \tilde{l}_{11} \\ \vdots \\ \tilde{l}_{1p} \end{pmatrix}$. The purpose of the WPCA based on fuzzy clustering is to estimate $\tilde{\boldsymbol{l}}_1$ which maximizes the variance of $\tilde{\boldsymbol{z}}_1$ under the condition of $\tilde{\boldsymbol{l}}_1'\tilde{\boldsymbol{l}}_1 = 1$. The variance of $\tilde{\boldsymbol{z}}_1$ is:

$$V\{\tilde{\boldsymbol{z}}_1\} = V\{\tilde{X}\tilde{\boldsymbol{l}}_1\} = \tilde{\boldsymbol{l}}_1'\tilde{\Sigma}\tilde{\boldsymbol{l}}_1, \tag{2.14}$$

where, $\tilde{\Sigma} = \tilde{X}'\tilde{X}$. Using the Lagrange's method of indeterminate multiplier, the following condition is needed for which $\tilde{\boldsymbol{l}}_1$ has a non-trivial solution.

$$|\tilde{\Sigma} - \tilde{\lambda}I| = 0, \tag{2.15}$$

where $\tilde{\lambda}$ is an indeterminate multiplier and I is a unit matrix. Equation (2.15) is the characteristic equation, so $\tilde{\lambda}$ is obtained as an eigen-value of $\tilde{\Sigma}$. Using equation (2.15) and the condition $\tilde{\boldsymbol{l}}_1'\tilde{\boldsymbol{l}}_1 = 1$, we obtain:

$$\tilde{\boldsymbol{l}}_1'\Sigma\tilde{\boldsymbol{l}}_1 = \tilde{\lambda}\tilde{\boldsymbol{l}}_1'\tilde{\boldsymbol{l}}_1 = \tilde{\lambda}. \tag{2.16}$$

From equations (2.14) and (2.16), $V\{\tilde{\boldsymbol{z}}_1\} = \tilde{\lambda}$. So, $\tilde{\boldsymbol{l}}_1$ is determined as the corresponding eigen-vector for the maximum eigen-value of $\tilde{\Sigma}$. In order to obtain the second principal component $\tilde{\boldsymbol{z}}_2$ for $\tilde{X}$, we define the linear combination:

$$\tilde{\boldsymbol{z}}_2 = \tilde{X}\tilde{\boldsymbol{l}}_2, \tag{2.17}$$

where $\tilde{\boldsymbol{l}}_2 = \begin{pmatrix} \tilde{l}_{21} \\ \vdots \\ \tilde{l}_{2p} \end{pmatrix}$. We need to estimate $\tilde{\boldsymbol{l}}_2$ which maximizes the variance of $\tilde{\boldsymbol{z}}_2$ under the condition of $\tilde{\boldsymbol{l}}_2'\tilde{\boldsymbol{l}}_2 = 1$ and covariance of $\tilde{\boldsymbol{z}}_1$ and $\tilde{\boldsymbol{z}}_2$ is 0, that is, $\tilde{\boldsymbol{z}}_1$ and $\tilde{\boldsymbol{z}}_2$ are mutually uncorrelated. This is represented as follows:

$$\mathrm{cov}\{\tilde{\boldsymbol{z}}_1, \tilde{\boldsymbol{z}}_2\} = \mathrm{cov}\{X\tilde{\boldsymbol{l}}_1, \tilde{X}\tilde{\boldsymbol{l}}_2\} = \tilde{\boldsymbol{l}}_1'\mathrm{cov}\{\tilde{X}, \tilde{X}\}\tilde{\boldsymbol{l}}_2 = \tilde{\boldsymbol{l}}_1'\tilde{\Sigma}\tilde{\boldsymbol{l}}_2 = \tilde{\lambda}\tilde{\boldsymbol{l}}_1'\tilde{\boldsymbol{l}}_2 = 0, \tag{2.18}$$

where $\mathrm{cov}\{\tilde{\boldsymbol{z}}_1, \tilde{\boldsymbol{z}}_2\}$ shows covariance between $\tilde{\boldsymbol{z}}_1$ and $\tilde{\boldsymbol{z}}_2$. We can obtain equation (2.18) from equation (2.16) and estimate $\tilde{\boldsymbol{l}}_2$ which satisfy $(\tilde{\Sigma} - \tilde{\lambda}I)\tilde{\boldsymbol{l}}_2 = 0$ under the conditions $\tilde{\boldsymbol{l}}_2'\tilde{\boldsymbol{l}}_2 = 1$ and $\tilde{\boldsymbol{l}}_1'\tilde{\boldsymbol{l}}_2 = 0$. From the fact that the eigen-vectors corresponding to the different eigen-values are mutually orthogonal, $\tilde{\boldsymbol{l}}_2$ is found as the eigen-vector corresponding to the second largest eigen-value of $\tilde{\Sigma}$. If we get s $(s \leq p)$ different eigen-values of $\tilde{\Sigma}$, according to the above, we find the s principal components for $\tilde{X}$.

The proportion of α-th principal component for WPCA based on fuzzy clustering, $\tilde{P}_\alpha$, is proposed as follows:

$$\tilde{P}_\alpha = \frac{\tilde{\lambda}_\alpha}{\mathrm{tr}(\tilde{\Sigma})}, \tag{2.19}$$

where $V\{\tilde{\boldsymbol{z}}_\alpha\} = \tilde{\lambda}_\alpha$ and $\sum_{\alpha=1}^{p} \tilde{\lambda}_\alpha = \mathrm{tr}(\tilde{\Sigma})$. The accumulated proportion until s-th principal component for WPCA based on fuzzy clustering is defined as:

$$\tilde{P} = \sum_{\alpha=1}^{s} \tilde{P}_\alpha. \tag{2.20}$$

From equations (2.19) and (2.20), $0 < \tilde{P} \leq 1$. The factor loading $\tilde{r}_{\alpha,j}$ between the α-th principal component $\tilde{\boldsymbol{z}}_\alpha$ and the j-th variable $\tilde{\boldsymbol{x}}_j$ is proposed as the following:

$$\tilde{r}_{\alpha,j} = \frac{\mathrm{cov}\{\tilde{\boldsymbol{z}}_\alpha, \tilde{\boldsymbol{x}}_j\}}{\sqrt{V\{\tilde{\boldsymbol{z}}_\alpha\}V\{\tilde{\boldsymbol{x}}_j\}}} = \frac{\sqrt{\tilde{\lambda}_\alpha}\tilde{\boldsymbol{l}}_{\alpha j}}{\sqrt{\tilde{\sigma}_{jj}}}, \tag{2.21}$$

where $\tilde{\sigma}_{jj}$ is the variance of $\tilde{\boldsymbol{x}}_j$.

2.4 Weighted Principal Component Analyses for Interval-Valued Data

This section describes Weighted Principal Component Analyses (WPCA) for interval-valued data using the result of fuzzy clustering.

In data analysis, there is an area where we faithfully represent the realistic and complex situation of data. The data can also be analyzed directly in order to avoid risk of obtaining the wrong result from the observations. Recently, symbolic data analysis [6], [29] has been proposed as one method of analysis. It has generated tremendous interest among a number of researchers. Conventional data analysis usually uses single quantitative or categorical data. In symbolic data analysis, multivalued quantitative or multivalued categorical data, interval-valued data, and a set of values with associated weights (modal data) are also used. We focus here on interval-valued data. For example, if we observe the situation of how long people watch TV per day and ask a subject, "How long do you watch TV per day?". If one subject answers 4.5 hours, then this observation has a risk that it is not a realistic situation. An answer that is from 3 to 5 hours is more realistic. So, such an observation should be treated as interval-valued data.

In PCA for interval-valued data, the vertices method and the centers method have been proposed [6]. The proposed method in this section is an extension for the centers method. In the centers method, the arithmetic mean of minimum and maximum values is calculated for each interval-valued data and the singularized mean values are applied to conventional PCA. Here we assume the homogeneous weights for both minimum and maximum values. That is, we assume the interval-valued data under a uniform situation. This assumption is not always adaptable. For example, in the former example, the subject watched TV eight times for close to 5 hours, otherwise, twice for close on 3 hours out of ten. In this case, the weights under the above assumption are not adjusted for a realistic situation.

If we know the frequency distribution of eight times and two times in the above example, we can then treat the frequency as a weight directly and treat the data as modal data. Such a case of interval-valued data seldom occurs.

In order to avoid the above risk of wrong assumption on the weight of the interval-valued data, we propose methods to estimate the weights for minimum and maximum values of each object. Firstly, we use the results of fuzzy clustering as obtained in [5], [36]. In addition a weighted principal component analysis based on the estimated weights is used [48], [56]. We then decompose the interval-valued data into two sets of data. One consists of the minimum values and the other consists of the maximum values. Next we calculate the weights for both data sets of the minimum and maximum values, respectively. A weighted principal component analysis for interval-valued data is proposed by use of a weighted linear combination [48], [56]. This method is a simple interaction between the minimum and maximum parts. Since we independently calculate the weights for both data sets of the minimum and maximum values, the merit of this method is that we do not need to assume the same number of clusters for both data sets. That is, we can use the adaptable number of clusters for both of the data sets.

However, in this method, we cannot guarantee that the assumption of weighted linear combination is justified or not. In order to solve this problem, another weighted principal component analysis for interval-valued data is also presented [52]. In this analysis, the weights are estimated as the results of fuzzy clustering for the minimum and the maximum parts in the given interval-valued data having unique clusters for the two parts. From this, we create a single weighted matrix without any assumptions when combining the two parts.

With this method we lose the merit of assuming the different number of clusters for the two parts. However, using the property of the uniqueness of the clusters over the two parts we can obtain two comparable principal components for the minimum and maximum parts of the interval-valued data. From this comparability, we can obtain the dynamic movement of the results from minimum to maximum.

2.4.1 WPCA for Interval-Valued Data based on Weighted Linear Combination

Suppose that the observed interval-valued data which are values of n objects with respect to p variables are denoted by the following:

$$X = ([\underline{x}_{ia}, \overline{x}_{ia}]), \quad i = 1, \ldots, n, \ a = 1, \ldots, p, \tag{2.22}$$

where $[\underline{x}_{ia}, \overline{x}_{ia}]$ shows the interval-valued data of the i-th object with respect to a variable a which has the minimum value $\underline{x}_{ia}$ and the maximum value $\overline{x}_{ia}$.

The data X is decomposed into two data sets $\underline{X}$ and $\overline{X}$ for the minimum values and the maximum values as follows:

$$\underline{X} = (\underline{x}_{ia}), \ \overline{X} = (\overline{x}_{ia}), \quad i = 1, \ldots, n, \ a = 1, \ldots, p. \tag{2.23}$$

For the analysis of interval-valued data, the question of how to combine $\underline{X}$ and $\overline{X}$ is an important problem. WPCA [48] assumes the following weighted linear combination of $\underline{X}$ and $\overline{X}$:

$$\hat{X} \equiv \underline{WX} + \overline{WX}. \tag{2.24}$$

$\underline{W}$ and $\overline{W}$ are diagonal matrixes defined as follows:

$$\underline{W} = \begin{pmatrix} \prod_{k=1}^{K} \underline{u}_{1k}^{-1} & \cdots & 0 \\ \vdots & \ddots & \vdots \\ 0 & \cdots & \prod_{k=1}^{K} \underline{u}_{nk}^{-1} \end{pmatrix} \equiv \begin{pmatrix} \underline{w}_1 & \cdots & 0 \\ \vdots & \ddots & \vdots \\ 0 & \cdots & \underline{w}_n \end{pmatrix}, \tag{2.25}$$

$$\overline{W} = \begin{pmatrix} \prod_{k'=1}^{K'} \overline{u}_{1k'}^{-1} & \cdots & 0 \\ \vdots & \ddots & \vdots \\ 0 & \cdots & \prod_{k'=1}^{K'} \overline{u}_{nk'}^{-1} \end{pmatrix} \equiv \begin{pmatrix} \overline{w}_1 & \cdots & 0 \\ \vdots & \ddots & \vdots \\ 0 & \cdots & \overline{w}_n \end{pmatrix}, \tag{2.26}$$

where $\underline{U} = (\underline{u}_{ik})$ and $\overline{U} = (\overline{u}_{ik'})$, $i = 1, \ldots, n$, $k = 1, \ldots, K$, $k' = 1, \ldots, K'$ are the results of fuzzy clustering obtained for $\underline{X}$ and $\overline{X}$, respectively. That is, $\underline{u}_{ik}$ shows the degree of belongingness of an object i to a cluster k for $\underline{X}$ and $\overline{u}_{ik'}$ shows the degree of belongingness of an object i to a cluster k' for $\overline{X}$. Notice that $\underline{u}_{ik}$ and $\overline{u}_{ik'}$ are obtained by separately classifying the data $\underline{X}$ and $\overline{X}$. Here $\underline{u}_{ik}$ and $\overline{u}_{ik'}$ are assumed to satisfy the following conditions:

$$\underline{u}_{ik},\ \overline{u}_{ik'} \in (0,1),\ \forall i, k, k';\ \sum_{k=1}^{K} \underline{u}_{ik} = \sum_{k'=1}^{K'} \overline{u}_{ik'} = 1,\ \forall i, \tag{2.27}$$

where K is the number of clusters for $\underline{X}$ and K' is the number of clusters for $\overline{X}$. This is similar to the weights shown in equation (2.10), using a property of the algebraic product and the condition of the degree of belongingness shown in equation (2.27). These weights show the clustering situations in the minimum and maximum parts of the interval-valued data. That is, the weight $\underline{w}_i$ shows that if the belonging status of the object i to the clusters in the minimum part is clearer, then the weight becomes larger. If the belonging status of the object i is less obvious in the minimum part, the classification structure of the data which consists of minimum values is close to uniformity, then the weight becomes small.

Therefore, the following equation

$$\underline{WX} = \begin{pmatrix} \underline{w}_1 & \cdots & 0 \\ \vdots & \ddots & \vdots \\ 0 & \cdots & \underline{w}_n \end{pmatrix} \begin{pmatrix} \underline{x}_{11} & \cdots & \underline{x}_{1p} \\ \vdots & \ddots & \vdots \\ \underline{x}_{n1} & \cdots & \underline{x}_{np} \end{pmatrix} = \begin{pmatrix} \underline{w}_1\underline{x}_{11} & \cdots & \underline{w}_1\underline{x}_{1p} \\ \vdots & \ddots & \vdots \\ \underline{w}_n\underline{x}_{n1} & \cdots & \underline{w}_n\underline{x}_{np} \end{pmatrix} \tag{2.28}$$

shows that clearer objects in the part of minimum values under the classification structure have larger values and that the objects which are vaguely situated under the classification structure are avoided. That is, those which do not have any significant relation to the classification structure for example objects that are treated as noise are avoided. For $\overline{WX}$ shown in equation (2.24), we can see the same relationships as that obtained above from condition (2.27). In equation (2.24), if $\underline{w}_1 = \cdots = \underline{w}_n = \overline{w}_1 = \cdots = \overline{w}_n = \frac{1}{2}$, then this method is the same as the centers method [6]. Therefore, this method is an extension of the centers method.

Suppose that $\hat{X}$ is centered with respect to p variables and $\hat{\boldsymbol{z}}_1$ is the first principal component of the transformed data $\hat{X}$. $\hat{\boldsymbol{z}}_1$ is defined as:

$$\hat{\boldsymbol{z}}_1 = \hat{X}\hat{\boldsymbol{l}}_1, \tag{2.29}$$

where $\hat{\boldsymbol{l}}_1 = \begin{pmatrix} \hat{l}_{11} \\ \vdots \\ \hat{l}_{1p} \end{pmatrix}$. The purpose of WPCA for interval-valued data based on fuzzy clustering is to estimate $\hat{\boldsymbol{l}}_1$ which maximizes the variance of $\hat{\boldsymbol{z}}_1$ under the condition of $\hat{\boldsymbol{l}}_1'\hat{\boldsymbol{l}}_1 = 1$. The estimate of $\hat{\boldsymbol{l}}_1$ is obtained as the corresponding eigen-vector for the maximum eigen-value of

$$\hat{\Sigma} = \hat{X}'\hat{X}. \tag{2.30}$$

In order to obtain the second principal component $\hat{\boldsymbol{z}}_2$ for $\hat{X}$, we define the following linear combination:

$$\hat{\boldsymbol{z}}_2 = \hat{X}\hat{\boldsymbol{l}}_2, \tag{2.31}$$

where $\hat{\boldsymbol{l}}_2 = \begin{pmatrix} \hat{l}_{21} \\ \vdots \\ \hat{l}_{2p} \end{pmatrix}$. $\hat{\boldsymbol{l}}_2$ is obtained as the eigen-vector corresponding to the second largest eigen-value of $\hat{\Sigma}$. If we get s $(s \leq p)$ different eigen-values of $\hat{\Sigma}$, according to the above, we find the s principal components for $\hat{X}$.

The proportion of the α-th principal component for WPCA for interval-valued data based on fuzzy clustering, $\hat{P}_\alpha$, is as follows:

$$\hat{P}_\alpha = \frac{\hat{\lambda}_\alpha}{\mathrm{tr}(\hat{\Sigma})}, \tag{2.32}$$

where $V\{\hat{\boldsymbol{z}}_\alpha\} = \hat{\lambda}_\alpha$ and $\sum_{\alpha=1}^{p} \hat{\lambda}_\alpha = \mathrm{tr}(\hat{\Sigma})$. The accumulated proportion until the s-th principal component for WPCA of the interval-valued data based on fuzzy clustering is defined as:

$$\hat{P} = \sum_{\alpha=1}^{s} \hat{P}_\alpha. \tag{2.33}$$

From equations (2.32) and (2.33), $0 < \hat{P} \leq 1$. The factor loading $\hat{r}_{\alpha,j}$ between the α-th principal component $\hat{\boldsymbol{z}}_\alpha$ and the j-th variable $\hat{\boldsymbol{x}}_j$ of $\hat{X}$ is proposed as follows:

$$\hat{r}_{\alpha,j} = \frac{\mathrm{cov}\{\hat{\boldsymbol{z}}_\alpha, \hat{\boldsymbol{x}}_j\}}{\sqrt{V\{\hat{\boldsymbol{z}}_\alpha\}V\{\hat{\boldsymbol{x}}_j\}}} = \frac{\sqrt{\hat{\lambda}_\alpha}\hat{l}_{\alpha j}}{\sqrt{\hat{\sigma}_{jj}}},$$

where $\hat{\sigma}_{jj}$ is variance of $\hat{\boldsymbol{x}}_j$.

We define the following two factor loadings $\underline{r}_{\alpha,j}$ and $\overline{r}_{\alpha,j}$ between the α-th principal component $\hat{\boldsymbol{z}}_\alpha$ and the j-th variable $\underline{\boldsymbol{x}}_j$ of $\underline{X}$ and the j-th variable $\overline{\boldsymbol{x}}_j$ of $\overline{X}$, respectively:

$$\underline{r}_{\alpha,j} = \frac{\mathrm{cov}\{\hat{\boldsymbol{z}}_\alpha, \underline{\boldsymbol{x}}_j\}}{\sqrt{V\{\hat{\boldsymbol{z}}_\alpha\}V\{\underline{\boldsymbol{x}}_j\}}}, \quad \overline{r}_{\alpha,j} = \frac{\mathrm{cov}\{\hat{\boldsymbol{z}}_\alpha, \overline{\boldsymbol{x}}_j\}}{\sqrt{V\{\hat{\boldsymbol{z}}_\alpha\}V\{\overline{\boldsymbol{x}}_j\}}}.$$

If $\hat{X}$ is a conventional data matrix, then the method is essentially the same as conventional PCA. However, the meaning of the variance shown in equation (2.30), which are the criterion in PCA, is different from the conventional method. In equation (2.30),

$$\begin{aligned}\hat{\Sigma} &= \hat{X}'\hat{X} \\ &= (\underline{WX} + \overline{WX})'(\underline{WX} + \overline{WX}) \\ &= \underline{X}'\underline{W}^2\underline{X} + \underline{X}'\underline{W}'\overline{WX} + \overline{X}'\overline{W}'\underline{WX} + \overline{X}'\overline{W}^2\overline{X},\end{aligned} \tag{2.34}$$

where,

$$\underline{W}^2 = \begin{pmatrix} \underline{w}_1^2 & \cdots & 0 \\ \vdots & \ddots & \vdots \\ 0 & \cdots & \underline{w}_n^2 \end{pmatrix}, \quad \overline{W}^2 = \begin{pmatrix} \overline{w}_1^2 & \cdots & 0 \\ \vdots & \ddots & \vdots \\ 0 & \cdots & \overline{w}_n^2 \end{pmatrix}.$$

In equation (2.34), the second and third terms show the interaction between $\underline{X}$ and $\overline{X}$, because

$$\underline{X}'\underline{W}'\overline{W}\overline{X} = (\underline{X}^*)'\overline{X}^*, \quad \overline{X}'\overline{W}'\underline{W}\underline{X} = (\overline{X}^*)'\underline{X}^*,$$

where,

$$\underline{X}^* \equiv \tilde{W}\underline{X}, \ \overline{X}^* \equiv \tilde{W}\overline{X}, \text{ and } \tilde{W} \equiv \begin{pmatrix} \sqrt{\underline{w}_1\overline{w}_1} & \cdots & 0 \\ \vdots & \ddots & \vdots \\ 0 & \cdots & \sqrt{\underline{w}_n\overline{w}_n} \end{pmatrix},$$

usually

$$(\underline{X}^*)'\overline{X}^* \neq (\overline{X}^*)'\underline{X}^*,$$

due to the asymmetry of $(\underline{X}^*)'\overline{X}^*$. This is a significant property of considering the weighted combination of $\underline{X}$ and $\overline{X}$ and a large difference from the independent treatment of $\underline{X}$ and $\overline{X}$.

2.4.2 WPCA for Interval-Valued Data Considering Uniqueness of Clusters

Using $\underline{X}$ and $\overline{X}$, we create a $2n \times p$ super matrix G as follows:

$$G \equiv \begin{pmatrix} \underline{X} \\ \overline{X} \end{pmatrix}. \qquad (2.35)$$

We apply a fuzzy clustering to G based on the idea of the dynamic clustering for temporal data and obtain the result of the degree of belongingness as follows:

$$U \equiv \begin{pmatrix} \underline{U}^* \\ \overline{U}^* \end{pmatrix}, \qquad (2.36)$$

$$\underline{U}^* = \begin{pmatrix} \underline{u}_{11}^* & \cdots & \underline{u}_{1K}^* \\ \vdots & \ddots & \vdots \\ \underline{u}_{n1}^* & \cdots & \underline{u}_{nK}^* \end{pmatrix}, \quad \overline{U}^* = \begin{pmatrix} \overline{u}_{11}^* & \cdots & \overline{u}_{1K}^* \\ \vdots & \ddots & \vdots \\ \overline{u}_{n1}^* & \cdots & \overline{u}_{nK}^* \end{pmatrix}.$$

Notice that $\underline{u}_{ik}^*$ and $\overline{u}_{ik}^*$ are obtained over the same clusters $C_1, \ldots, C_K$ for both $\underline{X}$ and $\overline{X}$. Compared with this, $\underline{u}_{ik}$ and $\overline{u}_{ik'}$ in equation (2.27) do not guarantee that they are obtained under the same clusters for both $\underline{X}$ and $\overline{X}$, even if we assume $K = K'$ in equation (2.27).
Then we create a matrix $\tilde{G}$ as follows:

$$\tilde{G} \equiv \begin{pmatrix} \underline{W}^* \underline{X} \\ \overline{W}^* \overline{X} \end{pmatrix}, \tag{2.37}$$

where

$$\underline{W}^* = \begin{pmatrix} \prod_{k=1}^{K} \underline{u}_{1k}^{*^{-1}} & \cdots & 0 \\ \vdots & \ddots & \vdots \\ 0 & \cdots & \prod_{k=1}^{K} \underline{u}_{nk}^{*^{-1}} \end{pmatrix} \equiv \begin{pmatrix} \underline{w}_1^* & \cdots & 0 \\ \vdots & \ddots & \vdots \\ 0 & \cdots & \underline{w}_n^* \end{pmatrix}, \tag{2.38}$$

$$\overline{W}^* = \begin{pmatrix} \prod_{k=1}^{K} \overline{u}_{1k}^{*^{-1}} & \cdots & 0 \\ \vdots & \ddots & \vdots \\ 0 & \cdots & \prod_{k=1}^{K} \overline{u}_{nk}^{*^{-1}} \end{pmatrix} \equiv \begin{pmatrix} \overline{w}_1^* & \cdots & 0 \\ \vdots & \ddots & \vdots \\ 0 & \cdots & \overline{w}_n^* \end{pmatrix}. \tag{2.39}$$

Suppose $\tilde{\tilde{G}}$ be a centered matrix with respect to p variables for $\tilde{G}$ and $\tilde{\tilde{\boldsymbol{z}}}_1$ is the first principal component of $\tilde{\tilde{G}}$. $\tilde{\tilde{\boldsymbol{z}}}_1$ is defined as:

$$\tilde{\tilde{\boldsymbol{z}}}_1 = \tilde{\tilde{G}} \tilde{\tilde{\boldsymbol{l}}}_1, \tag{2.40}$$

where $\tilde{\tilde{\boldsymbol{l}}}_1 = \begin{pmatrix} \tilde{\tilde{l}}_{11} \\ \vdots \\ \tilde{\tilde{l}}_{1p} \end{pmatrix}$. $\tilde{\tilde{\boldsymbol{z}}}_1$ is represented as:

$$\tilde{\tilde{\boldsymbol{z}}}_1 = (\tilde{\tilde{z}}_{11}, \cdots, \tilde{\tilde{z}}_{1n}, \tilde{\tilde{z}}_{1,n+1}, \ldots, \tilde{\tilde{z}}_{1,2n})'. \tag{2.41}$$

In equation (2.41), from the first row to the n-th row show the elements of the first principal component for the minimum part of the data. The elements from the $(n+1)$-th row to $2n$-th row show the elements of the first principal component for the maximum part of the interval-valued data. By the use of the same clusters for both the minimum and maximum parts, we can compare these results of the two parts for the first principal component. This is a clear difference when compared to conventional WPCA, because when using equation (2.29), we can obtain only the single result for the first principal component in conventional WPCA.

The purpose of WPCA for interval-valued data based on unique fuzzy clusters is to estimate $\tilde{\tilde{l}}_1$ which maximizes the variance of $\tilde{\tilde{z}}_1$ under the condition of $\tilde{\tilde{l}}_1'\tilde{\tilde{l}}_1 = 1$. The estimate of $\tilde{\tilde{l}}_1$ is obtained as the corresponding eigen-vector for the maximum eigen-value of $\tilde{\tilde{\Sigma}} = \tilde{\tilde{G}}'\tilde{\tilde{G}}$.

By using the second largest eigen-value of $\tilde{\tilde{\Sigma}}$, we can obtain the second principal component and if we get s $(s \leq p)$ different eigen-values of $\tilde{\tilde{\Sigma}}$, according to the above, we find the s principal components for $\tilde{\tilde{G}}$.

The proportion of α-th principal component for the proposed WPCA for interval-valued data based on unique fuzzy clusters, $\tilde{\tilde{P}}_\alpha$, is as follows:

$$\tilde{\tilde{P}}_\alpha = \frac{\tilde{\tilde{\lambda}}_\alpha}{\mathrm{tr}(\tilde{\tilde{\Sigma}})}, \tag{2.42}$$

where $V\{\tilde{\tilde{z}}_\alpha\} = \tilde{\tilde{\lambda}}_\alpha$ and $\sum_{\alpha=1}^{p} \tilde{\tilde{\lambda}}_\alpha = \mathrm{tr}(\tilde{\tilde{\Sigma}})$. The accumulated proportion until s-th principal component for the WPCA of interval-valued data based on unique fuzzy clusters is defined as:

$$\tilde{\tilde{P}} = \sum_{\alpha=1}^{s} \tilde{\tilde{P}}_\alpha. \tag{2.43}$$

From equations (2.42) and (2.43), $0 < \tilde{\tilde{P}} \leq 1$.

The considerable difference between the result of our former research shown in section 2.4.1 [48] and this research is that (1) we can obtain the degree of belongingness of $\underline{X}$ and $\overline{X}$ under the same clusters. Then we can compare the results of WPCA for $\underline{X}$ and $\overline{X}$ and (2) we can avoid the risk of a wrong assumption for the combination of $\underline{X}$ and $\overline{X}$.

2.5 Numerical Examples for Weighted Principal Component Analysis

2.5.1 Numerical Examples of WPCA for Single Valued Data

Let us consider the Fisher iris data [13]. The data consists of 150 samples of iris flowers with respect to four variables, sepal length, sepal width, petal length, and petal width. The samples are observed from three kinds of iris flowers, iris setosa, iris versicolor, and iris virginica.

Figure 2.1 shows the result of conventional PCA for the iris data [64]. In this figure, the abscissa shows the values of the first principal component shown in equation (2.2) and the ordinate shows the values of the second principal component shown in equation (2.6). "a" means iris setosa, "b" means iris versicolor, and "c" is iris virginica. From this figure, we can see iris setosa (the symbol is "a") is clearly separated from iris versicolor and iris virginica (the symbols are "b" and "c").

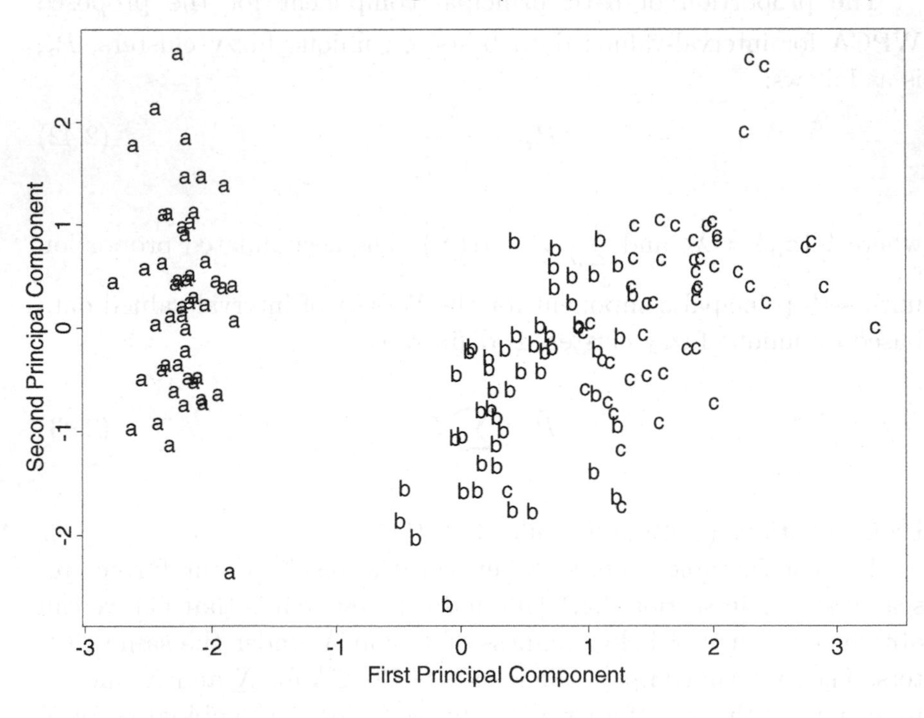

Fig. 2.1. Result of PCA for Iris Data

Figure 2.2 shows the proportion shown in equation (2.7) and accumulated proportion shown in equation (2.8) of the four components. In this figure, the abscissa shows each component and the ordinate shows the values of the proportion of each component. Also the value on the bar graph shows the accumulated proportion. From this figure, we can see that it can almost be explained by the first and the second principal components. However, we can not ignore the third principal component completely.

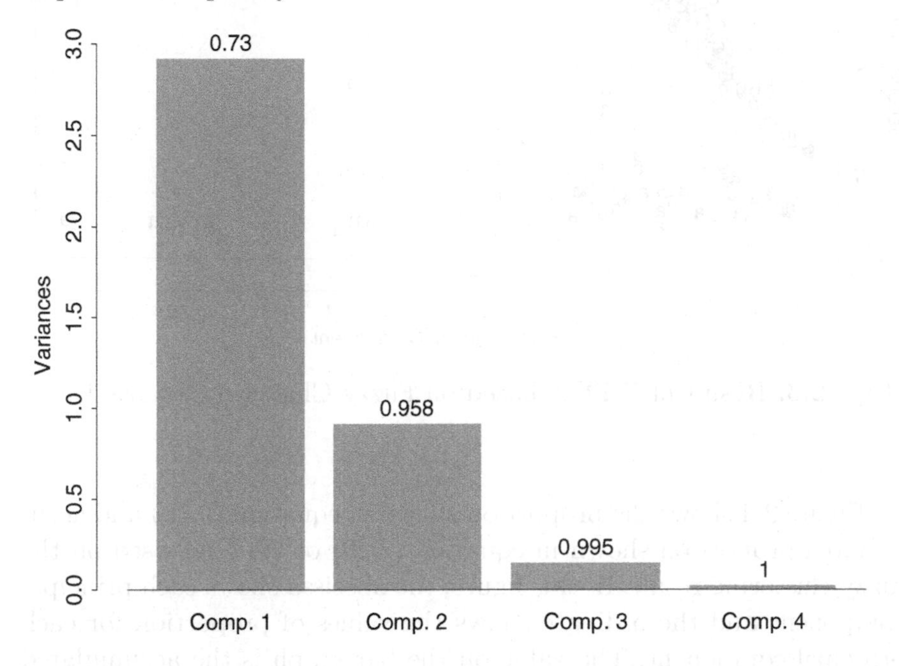

Fig. 2.2. Result of Proportion for PCA

Figure 2.3 shows the result of WPCA based on fuzzy clustering for the single valued data shown in section 2.3. As a weight W shown in equation (2.10), we use a fuzzy clustering result which is obtained by using the FANNY algorithm [27] whose objective function is equation (1.4) when $m = 2$. In figure 2.3, the abscissa shows the values of the first principal component shown in equation (2.13) and the ordinate is the values of the second principal component shown in equation (2.17). From this figure, we can see a clear distinction between iris setosa (the symbol is "a") and iris versicolor, iris virginica (the symbols are "b" and "c"), when comparing the result shown in figure 2.1.

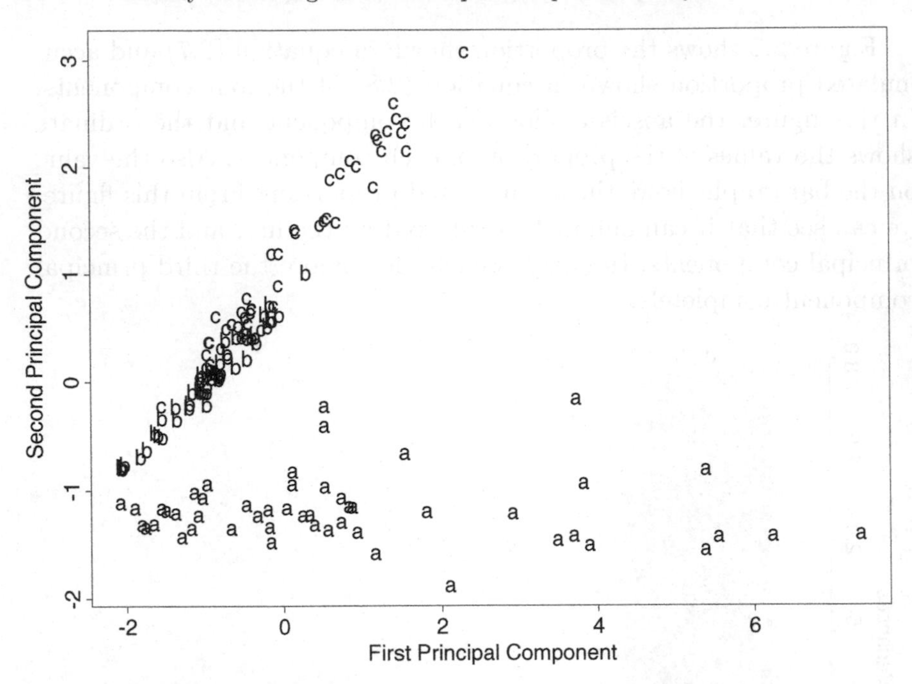

Fig. 2.3. Result of WPCA based on Fuzzy Clustering for Iris Data

Figure 2.4 shows the proportion shown in equation (2.19) and accumulated proportion shown in equation (2.20) of WPCA based on the fuzzy clustering result. In this figure, the abscissa shows each principal component and the ordinate shows the values of proportion for each principal component. The value on the bar graph is the accumulated proportion for each principal component.

By comparing figure 2.2 with figure 2.4, we can see the higher value of the accumulated proportion until the second principal component in WPCA based on fuzzy clustering (0.991) than the value of the accumulated proportion until the second principal component of PCA (0.958). This shows a higher discrimination ability with WPCA based on fuzzy clustering.

Moreover, we cannot see any significant meaning for the third principal component from the result of the proportion of WPCA based on fuzzy clustering shown in figure 2.4. So, we can avoid the noise of the data by the introduction of the weights based on the fuzzy clustering result.

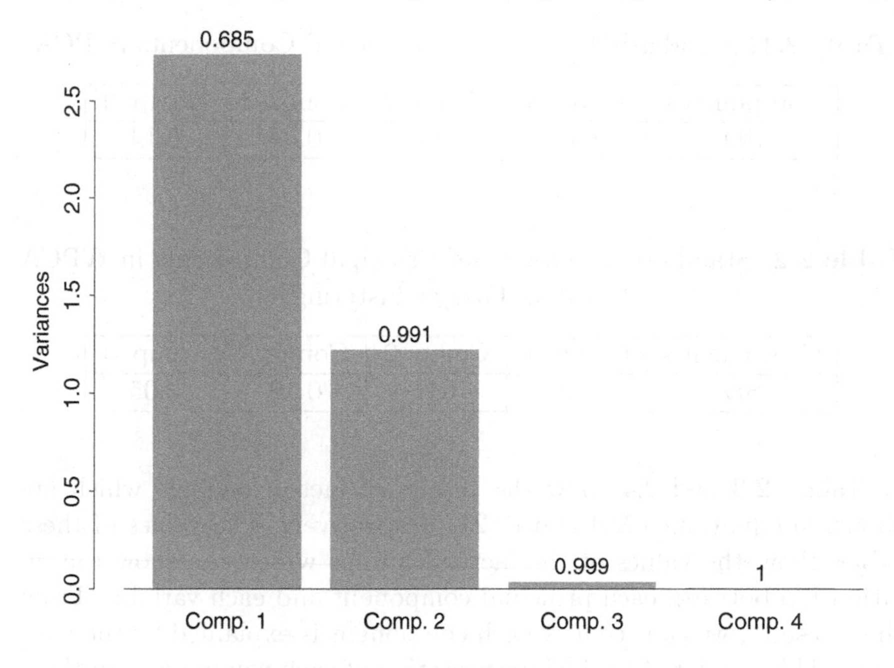

Fig. 2.4. Result of Proportion for WPCA based on Fuzzy Clustering

Tables 2.1 and 2.2 show the square root of λ_α shown in equation (2.5) and the square root of $\tilde{\lambda}_\alpha$ shown in equation (2.16), respectively. In these tables comp.1 shows the first principal component, comp.2 shows the second principal component, comp. 3 is the third principal component, and comp. 4 is the fourth principal component. SD shows standard deviation. From the comparison between the results of tables 2.1 and 2.2, we can see that the value for the second principal component in table 2.2 is higher than the value for the second principal component in table 2.1. We can also see smaller values for the third and the fourth principal components in table 2.2 when compared to the values for the third and the fourth principal components in table 2.1. From this, we can see that the WPCA based on fuzzy clustering has a higher capability to clearly capture the significance of the latent data structure. It tends to avoid the noise of the data, although the value for the first principal component becomes smaller in table 2.2.

Table 2.1. Standard Deviations for Principal Components in PCA

Components	Comp. 1	Comp. 2	Comp. 3	Comp. 4
SD	1.71	0.96	0.38	0.14

Table 2.2. Standard Deviations for Principal Components in WPCA based on Fuzzy Clustering

Components	Comp. 1	Comp. 2	Comp. 3	Comp. 4
SD	1.66	1.11	0.19	0.05

Tables 2.3 and 2.4 show the results of factor loadings which are shown in equations (2.9) and (2.21), respectively. The values of these tables show the values of the factor loadings which can show the relationship between each principal component and each variable. From these results, we can see how each component is explained by the variables. This is related to the interpretation of each component. In these tables, Sepal L. shows sepal length, Sepal W. shows sepal width, Petal L. shows petal length, and Petal W. is the petal width.

Table 2.3. Factor Loading in PCA

Variables	Comp. 1	Comp. 2	Comp. 3	Comp. 4
Sepal L.	0.89	0.36	0.28	0.04
Sepal W.	-0.46	0.88	-0.09	-0.02
Petal L.	0.99	0.02	-0.05	-0.12
Petal W.	0.96	0.06	-0.24	0.08

Table 2.4. Factor Loading in WPCA based on Fuzzy Clustering

Variables	Comp. 1	Comp. 2	Comp. 3	Comp. 4
Sepal L.	0.95	-0.36	-0.03	0.04
Sepal W.	0.90	-0.44	-0.07	-0.03
Petal L.	0.93	0.34	0.14	-0.01
Petal W.	0.45	0.89	-0.10	0.00

From the comparison between the results of tables 2.3 and 2.4, we can see quite different results. For example, in table 2.3, the first principal component is mainly explained by the variables, sepal length, petal length, and petal width. However, in table 2.4, the first principal component is explained by the variables, sepal length, sepal width, and petal length. The second principal component shown in table 2.3 indicates that the second principal component is represented by the strong relationship of the variable sepal width. However, in table 2.4, we can see a high correlation between the second principal component and the variable petal width.

From this, we can see that the principal components which have different meanings from the principal components obtained by using conventional PCA are captured by using WPCA based on fuzzy clustering. The ability to capture different latent factors of the data may be shown by introducing the classification structure of the data.

2.5.2 Numerical Examples of WPCA for Interval-Valued Data based on Weighted Linear Combination

We now use the oil data given in table 2.5 [25]. The data is observed as interval-valued data. We divide the data into minimum values and maximum values and create the two data matrixes shown in table 2.6.

Using the two data matrixes in a fuzzy clustering method named FANNY [27], the two results shown in table 2.7 are obtained. The number of clusters is assumed as 2. In table 2.7, each value shows the degree of belongingness of each oil to each cluster and each value is rounded. Using the results in table 2.7, the weights shown in table 2.8 are obtained for the diagonal parts of $\underline{W}$ and $\overline{W}$ in equations (2.25) and (2.26), respectively. Using the data shown in table 2.6 and the weights in table 2.8, $\hat{X}$ is calculated using equation (2.24).

Figure 2.5 shows the result of WPCA based on fuzzy clustering for $\hat{X}$. In this figure, the abscissa shows the values of the first principal component shown in equation (2.29), and the ordinate is the values of the second principal component shown in equation (2.31). From this figure, we can see the pairs, linseed oil and perilla oil, cottonseed oil and sesame oil, camellia oil and olive oil are similar to each other. It is known that linseed oil and perilla oil are used for paint, cottonseed oil and sesame oil for foods, camellia oil and olive oil for cosmetics, and so on. Moreover, we can see these similarities of the major fatty acids in table 2.5.

Table 2.5. Oil Data

Oils	Specific Gravity	Iodine Value	Saponification Value
Linseed Oil	0.930-0.935	170-204	118-196
Perilla Oil	0.930-0.935	192-208	188-197
Cottonseed Oil	0.916-0.918	99-113	189-198
Sesame Oil	0.920-0.926	104-116	187-193
Camellia Oil	0.916-0.917	80-82	189-193
Olive Oil	0.914-0.919	79-90	187-196
Beef Tallow	0.860-0.870	40-48	190-199
Hog Fat	0.858-0.864	53-77	190-202

Oils	Major Fatty Acids
Linseed Oil	L, Ln, O, P, M
Perilla Oil	L, Ln, O, P, S
Cottonseed Oil	L, O, P, M, S
Sesame Oil	L, O, P, S, A
Camellia Oil	L, O
Olive Oil	L, O, P, S
Beef Tallow	O, P, M, S, C
Hog Fat	L, O, P, M, S, Lu

L: linoleic acid, Ln: linolenic acid, O: oleic acid,
P: palmitic acid, M: myristic acid, S: searic acid,
A: arachic acid, C: capric acid, Lu: lauric acid

Table 2.6. Minimum Valued Data and Maximum Valued Data for Oil Data

Oils	$\underline{X}$ (Minimum)			$\overline{X}$ (Maximum)		
Linseed Oil	0.930	170	118	0.935	204	196
Perilla Oil	0.930	192	188	0.937	208	197
Cottonseed Oil	0.916	99	189	0.918	113	198
Sesame Oil	0.920	104	187	0.926	116	193
Camellia Oil	0.916	80	189	0.917	82	193
Olive Oil	0.914	79	187	0.919	90	196
Beef Tallow	0.860	40	190	0.870	48	199
Hog Fat	0.858	53	190	0.864	77	202

Table 2.7 Fuzzy Clustering Results for Oil Data

Oils	Grade for $\underline{X}$		Grade for $\overline{X}$	
	C_1	C_2	C_1	C_2
Linseed Oil	0.85	0.15	0.99	0.01
Perilla Oil	0.85	0.15	0.99	0.01
Cottonseed Oil	0.16	0.84	0.15	0.85
Sesame Oil	0.20	0.80	0.18	0.82
Camellia Oil	0.06	0.94	0.05	0.95
Olive Oil	0.06	0.94	0.05	0.95
Beef Tallow	0.17	0.83	0.15	0.85
Hog Fat	0.12	0.88	0.06	0.94

C_1: Cluster 1, C_2: Cluster 2

Table 2.8. Weights for Minimum and Maximum Data

Oils	Weights for $\underline{X}$	Weights for $\overline{X}$
Linseed Oil	7.74	95.90
Perilla Oil	7.66	86.37
Cottonseed Oil	7.54	7.77
Sesame Oil	6.16	6.90
Camellia Oil	18.11	21.58
Olive Oil	18.10	21.89
Beef Tallow	7.24	7.68
Hog Fat	9.54	17.43

Figures 2.6 (a) and (b) show the results for $\underline{X}$ and $\overline{X}$, respectively, that is, we use the data matrixes for minimum values and maximum values, independently. (c) shows the result of the centers method, where, we apply the following data $\check{X}$,

$$\check{X} \equiv \frac{\underline{X} + \overline{X}}{2}.$$

From figures 2.6 (a) and (c), we cannot see the similarity of linseed oil and perilla oil, and figure 2.6 (b) has a large dispersion when compared to the result of figure 2.5.

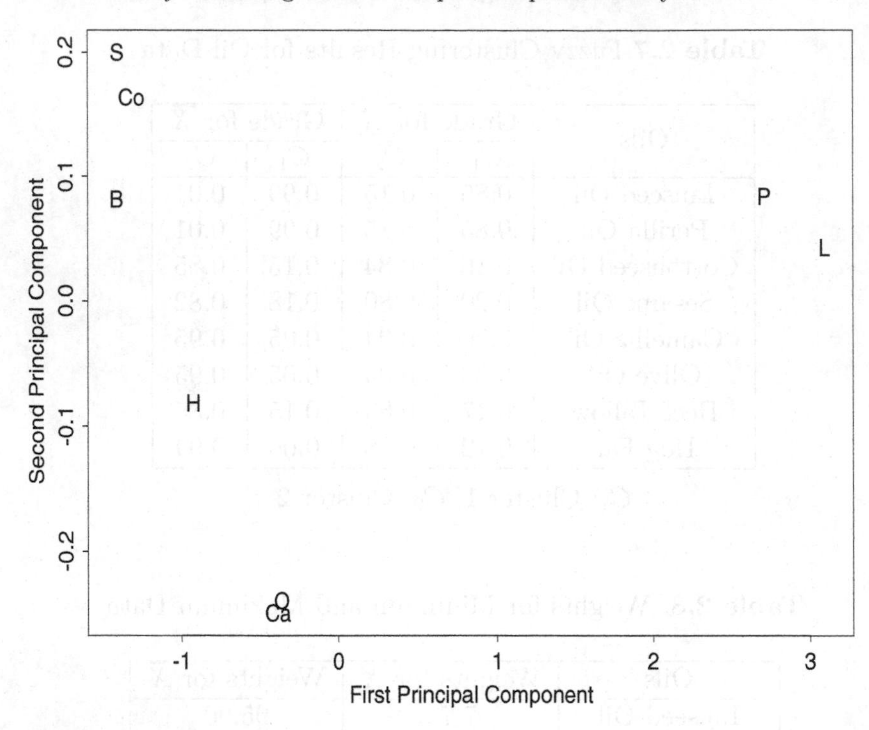

L: linseed oil, P: perilla oil, Co: cottonseed oil,
S: sesame oil, Ca: camellia oil, O: olive oil,
B: beef tallow, H: hog fat

Fig. 2.5. Result of WPCA when Applied to the Oil Data

Table 2.9 shows the comparison of the accumulated proportion shown in equation (2.33) of the first and the second principal components for $\hat{X}$, $\underline{X}$, $\overline{X}$, and $\check{X}$. From this, we obtain the best result for $\hat{X}$, that is the weighted method using the fuzzy clustering result has the best result for the proportion.

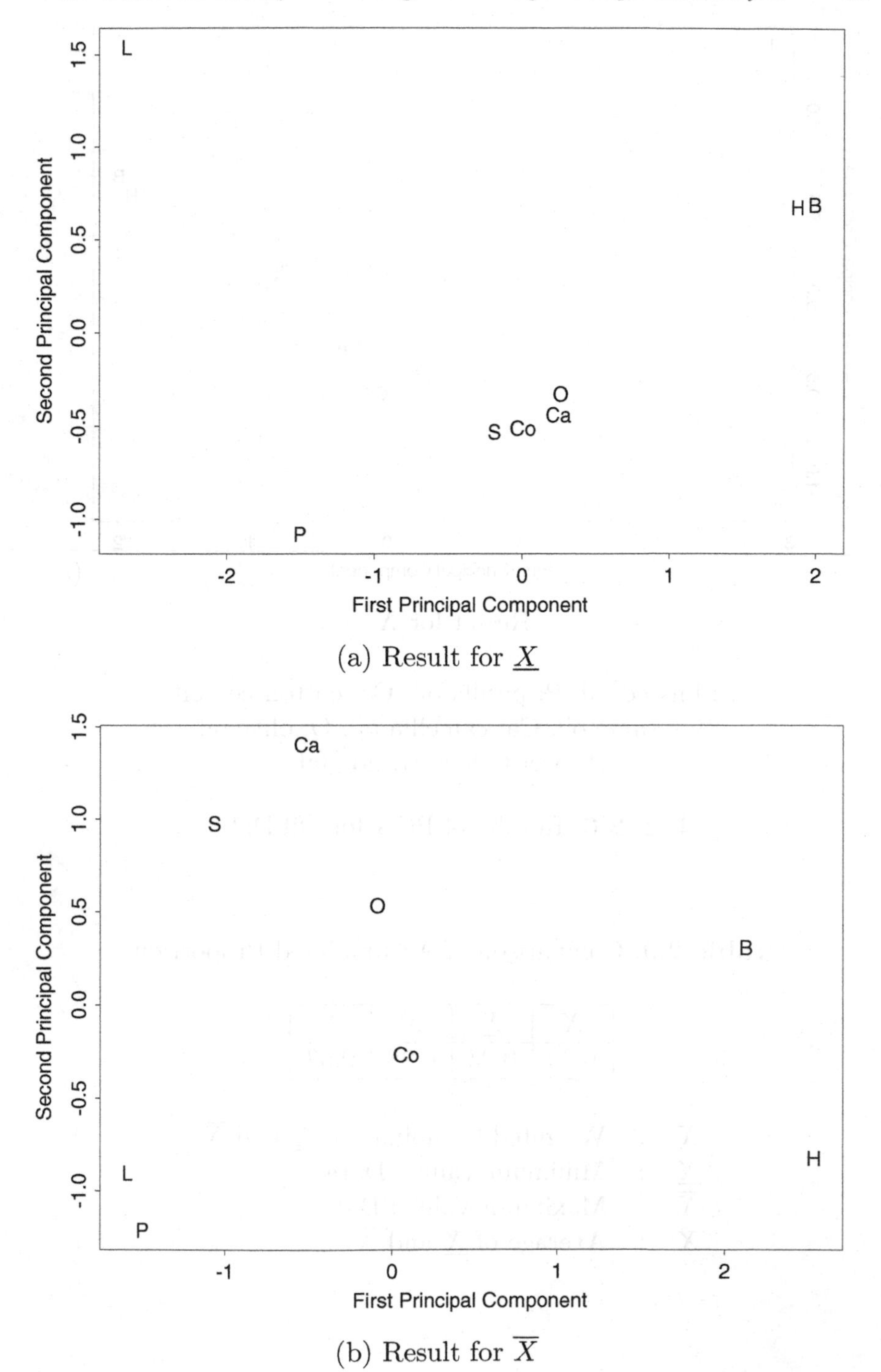

(a) Result for $\underline{X}$

(b) Result for $\overline{X}$

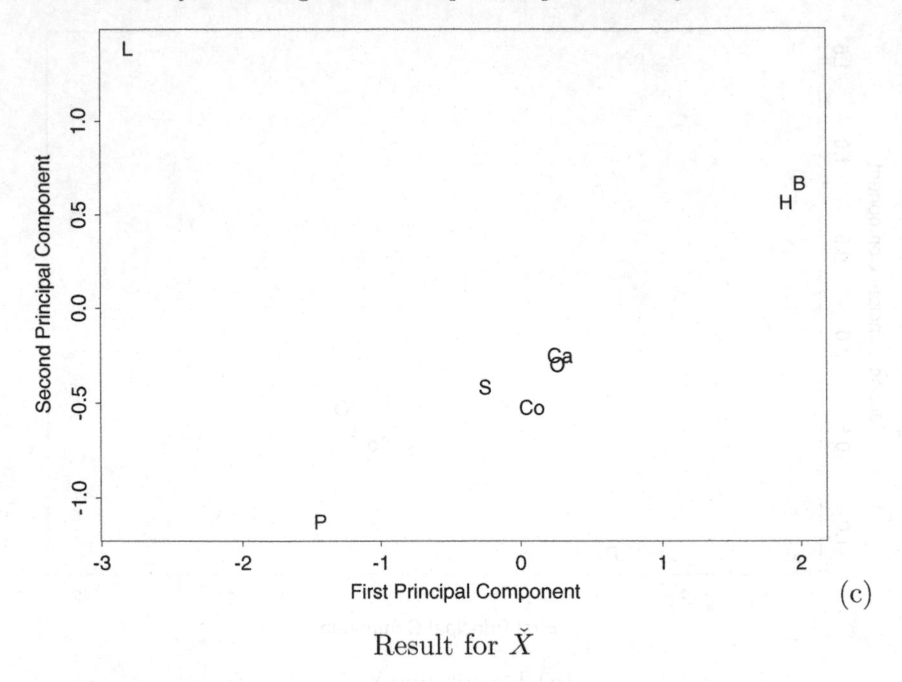

Result for $\check{X}$

L: linseed oil, P: perilla oil, Co: cottonseed oil,
S: sesame oil, Ca: camellia oil, O: olive oil,
B: beef tallow, H: hog fat

Fig. 2.6. Results of PCA for Oil Data

Table 2.9. Comparison of Accumulated Proportion

$\hat{X}$	$\underline{X}$	$\overline{X}$	$\check{X}$
0.99	0.93	0.98	0.93

$\hat{X}$: Weighted Combination $\underline{X}$ and $\overline{X}$
$\underline{X}$: Minimum Valued Data
$\overline{X}$: Maximum Valued Data
$\check{X}$: Average of $\underline{X}$ and $\overline{X}$

Figures 2.7 and 2.8 show the relationship between the weights in equations (2.25) and (2.26), and the classification structure in table 2.7. Figure 2.7 (a) shows the values of weights based on the minimum valued data, with respect to the fuzzy clustering result for clusters C_1 and C_2. We use the values of grades for $\underline{u}_{i1}^{-1}$ and $\underline{u}_{i2}^{-1}$, $(i = 1, \ldots, 8)$ in equation (2.25). This shows the values of $\underline{u}_{i1}^{-1}\underline{u}_{i2}^{-1}$ for all i, j as the surface in this figure. Figure 2.7 (b) is created in a similar way by using equation (2.26). Figures 2.8 (a) and (b) show the contours of figures 2.7 (a) and (b), respectively. In these figures, the symbols show each oil.

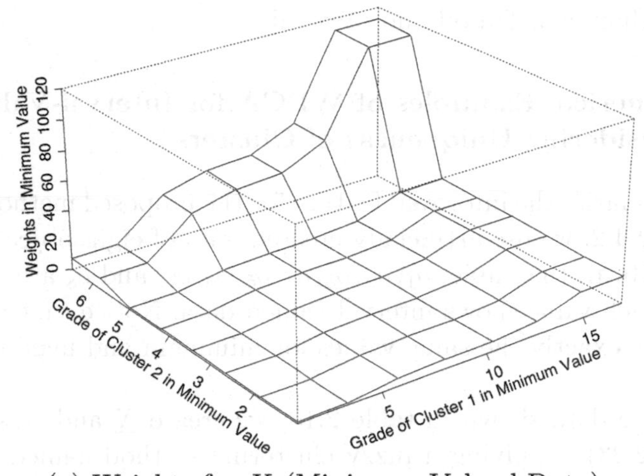

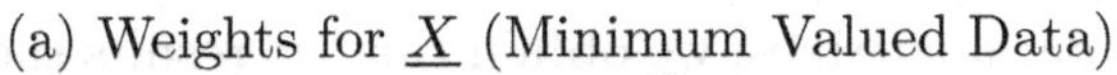

(a) Weights for $\underline{X}$ (Minimum Valued Data)

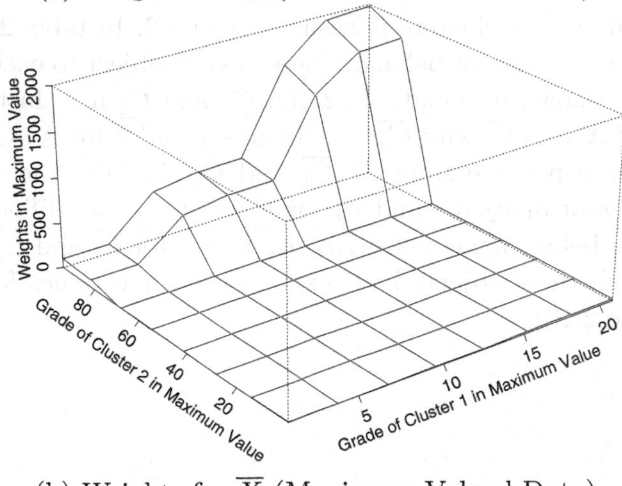

(b) Weights for $\overline{X}$ (Maximum Valued Data)

Fig. 2.7. Weights for Oil Data

In figures 2.7 and 2.8, we can see the clustering situation of these oils and the situation of the weights in equations (2.25) and (2.26), simultaneously. For example, in figure 2.8 (a), we can see the similarity of clustering situations of linseed oil and perilla oil, camellia oil and olive oil. Camellia oil and olive oil have large weights when compared to other oils. This can be seen in the higher contours for camellia oil and olive oil compared to the other oils. Moreover, from the comparison of figures 2.8 (a) and (b), the weights of linseed oil and perilla oil are smaller than camellia oil and olive oil in (a), however, in (b), we obtain the opposite situation, that is, the weights of linseed oil and perilla oil are larger than camellia oil and olive oil.

2.5.3 Numerical Examples of WPCA for Interval-Valued Data Considering Uniqueness of Clusters

In order to clarify the improved feature for the proposed method shown in section 2.4.2, we use artificially created data of eight objects shown in table 2.10. In this table, $o_1, o_2, o_3, o_4, o_5, o_6, o_7$, and o_8 are eight objects and each value shows interval-valued data. Notice that o_1, o_2, o_3, and o_4 have exactly the same values for minimum and maximum values.

Using the data shown in table 2.10, we create $\underline{X}$ and $\overline{X}$ shown in equation (2.23). Applying a fuzzy clustering method named FANNY [27] to the two data matrixes, we obtain the two results shown in table 2.11. The number of clusters is assumed to be 2. In table 2.11, each value shows the degree of belongingness of each object to each cluster, $\underline{u}_{ik}$ and $\overline{u}_{ik'}$, shown in equation (2.27). $\underline{C_1}$ and $\underline{C_2}$ are cluster 1 and cluster 2 for $\underline{X}$ and $\overline{C_1}$ and $\overline{C_2}$ are clusters 1 and 2 for $\overline{X}$. Generally, $\underline{C_1}$ and $\underline{C_2}$ are not the same as $\overline{C_1}$ and $\overline{C_2}$. In fact, the degree of belongingness of o_1, o_2, o_3, and o_4 for $\underline{C_1}$ and $\underline{C_2}$, are different from the degree of belongingness of o_1, o_2, o_3, and o_4 for $\overline{C_1}$ and $\overline{C_2}$ in table 2.11, although these objects have exactly the same values for $\underline{X}$ and $\overline{X}$. (see table 2.10)

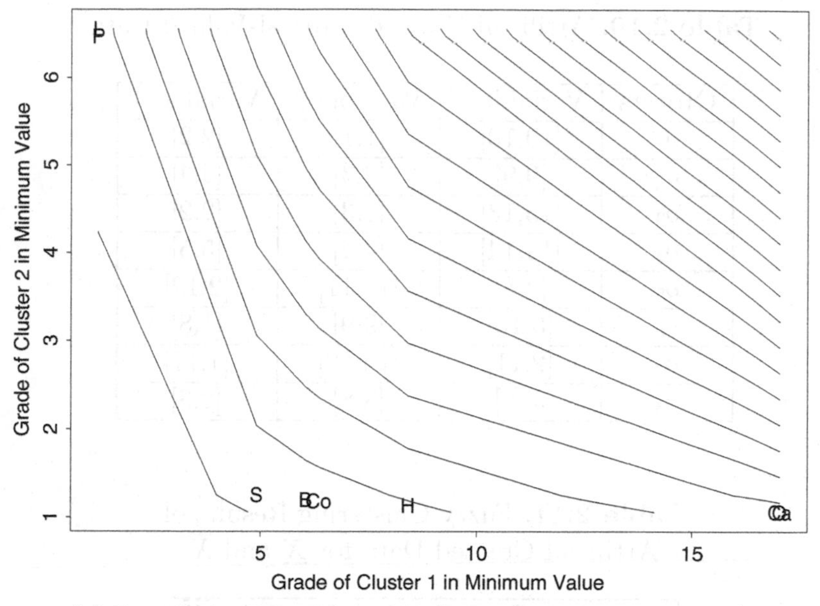

(a) Result for Weights in $\underline{X}$ (Minimum Valued Data)

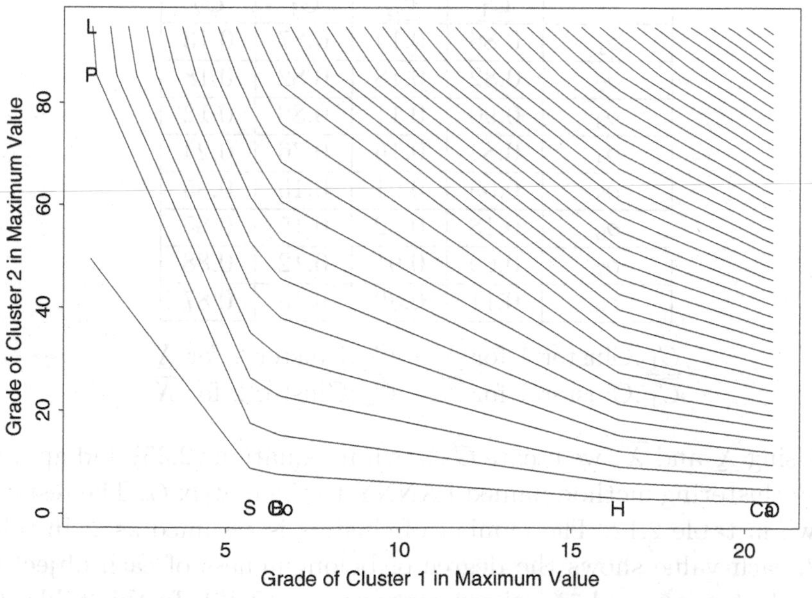

(b) Result for Weights in $\overline{X}$ (Maximum Valued Data)

L: linseed oil, P: perilla oil, Co: cottonseed oil, S: sesame oil, Ca: camellia oil, O: olive oil, B: beef tallow, H: hog fat

Fig. 2.8. Contour of Weights and Clustering Results

Table 2.10. Artificial Created Interval-Valued Data

Objects	Variable 1	Variable 2	Variable 3
o_1	[10,10]	[1,1]	[2,2]
o_2	[9,9]	[3,3]	[4,4]
o_3	[13,13]	[3,3]	[2,2]
o_4	[14,14]	[4,4]	[5,5]
o_5	[4,8]	[11,11]	[2,12]
o_6	[6,10]	[9,9]	[1,8]
o_7	[2,11]	[10,10]	[1,11]
o_8	[3,9]	[8,8]	[2,9]

Table 2.11. Fuzzy Clustering Results of Artificial Created Data for $\underline{X}$ and $\overline{X}$

Objects	Grade for $\underline{X}$		Grade for $\overline{X}$	
	$\underline{C_1}$	$\underline{C_2}$	$\overline{C_1}$	$\overline{C_2}$
o_1	0.86	0.14	0.87	0.13
o_2	0.82	0.18	0.82	0.18
o_3	0.90	0.11	0.88	0.12
o_4	0.84	0.16	0.76	0.24
o_5	0.09	0.91	0.15	0.85
o_6	0.18	0.82	0.15	0.85
o_7	0.09	0.91	0.12	0.88
o_8	0.11	0.89	0.13	0.87

$\underline{C_1}$:Cluster 1 for $\underline{X}$, $\underline{C_2}$:Cluster 2 for $\underline{X}$
$\overline{C_1}$:Cluster 1 for $\overline{X}$, $\overline{C_2}$:Cluster 2 for $\overline{X}$

Using $\underline{X}$ and $\overline{X}$, we create G shown in equation (2.35) and apply a fuzzy clustering method named FANNY to the matrix G. The result is shown in table 2.12. The number of clusters is assumed as 2. In table 2.12, each value shows the degree of belongingness of each object to each cluster, $\underline{u}^*_{ik}$ and $\overline{u}^*_{ik}$, shown in equation (2.36). In this table, C_1 and C_2 in $\underline{X}$ are exactly the same as C_1 and C_2 in $\overline{X}$. From this table, we can see that the degree of belongingness of objects o_1, o_2, o_3, and o_4 for C_1 and C_2 are the same for both $\underline{X}$ and $\overline{X}$.

Using the results of fuzzy clustering shown in table 2.11, we create weights $\underline{W}$, $\overline{W}$ in equations (2.25) and (2.26). Also using table 2.12, we create weights $\underline{W}^*$, and $\overline{W}^*$ in equations (2.38) and (2.39). We create two $\tilde{G}$ in equation (2.37) using the $\underline{W}$ with $\overline{W}$ and $\underline{W}^*$ with $\overline{W}^*$, respectively. We call these two $\tilde{G}$ as $\tilde{G}$, and $\tilde{G}^*$. By applying fuzzy clustering based WPCA to $\tilde{G}$ and $\tilde{G}^*$, we obtain the two results for $\tilde{G}$ and $\tilde{G}^*$, respectively. The results of the principal components obtained are shown in figures 2.9 and 2.10. Figure 2.9 shows the result of fuzzy clustering based WPCA for $\tilde{G}$. In this figure, the abscissa shows the values of the first principal component, and the ordinate is the values of the second principal component. Figure 2.10 shows the result of the first and the second principal components of fuzzy clustering based WPCA for $\tilde{G}^*$. In these figures, $1(L)$ to $8(L)$ show the results of the eight objects o_1 to o_8 for minimum values, $\underline{X}$, and $1(U)$ to $8(U)$ show the results of the eight objects o_1 to o_8 for maximum values, $\overline{X}$.

Table 2.12. Fuzzy Clustering Result of Artificial Created Data for G

Objects	Grade for $\underline{X}$		Grade for $\overline{X}$	
	C_1	C_2	C_1	C_2
o_1	0.80	0.20	0.80	0.20
o_2	0.76	0.24	0.76	0.24
o_3	0.83	0.17	0.83	0.17
o_4	0.77	0.23	0.77	0.23
o_5	0.23	0.77	0.33	0.67
o_6	0.29	0.71	0.34	0.66
o_7	0.27	0.73	0.37	0.63
o_8	0.27	0.73	0.35	0.65

C_1 :Cluster 1 for both $\underline{X}$ and $\overline{X}$
C_2 :Cluster 2 for both $\underline{X}$ and $\overline{X}$

From the comparison between figures 2.9 and 2.10, we can see that for objects o_1, o_2, o_3, and o_4, the results in $\underline{X}$ and $\overline{X}$ are exactly the same as those in figure 2.10. This is an adaptable result considering the structure of the given data shown in table 2.10. In figure 2.9, we can see that the results of objects o_1, o_2, o_3, and o_4 for $\underline{X}$ and $\overline{X}$ are different from each other. In this case, we cannot explain the data structure in table 2.10.

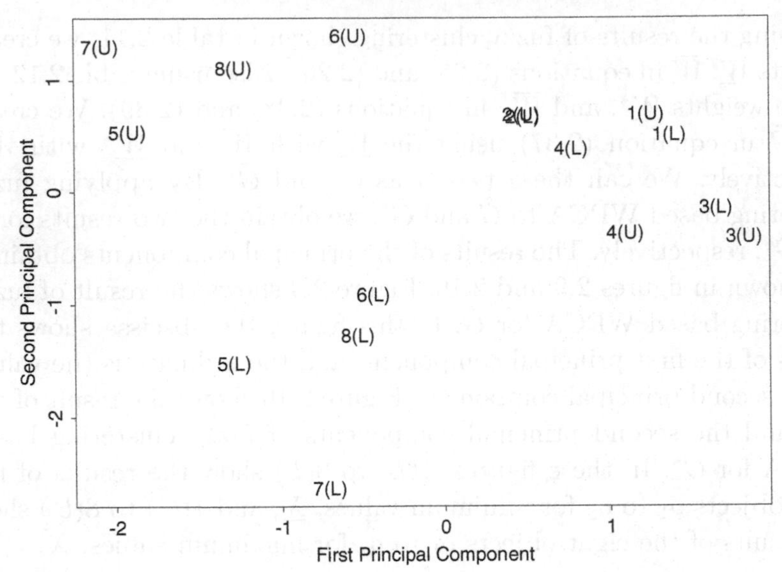

$1(L), 2(L), 3(L), 4(L), 5(L), 6(L), 7(L), 8(L)$: Eight Objects for $\underline{X}$
$1(U), 2(U), 3(U), 4(U), 5(U), 6(U), 7(U), 8(U)$: Eight Objects for $\overline{X}$

Fig. 2.9. Result of Fuzzy Clustering based WPCA for $\tilde{G}$

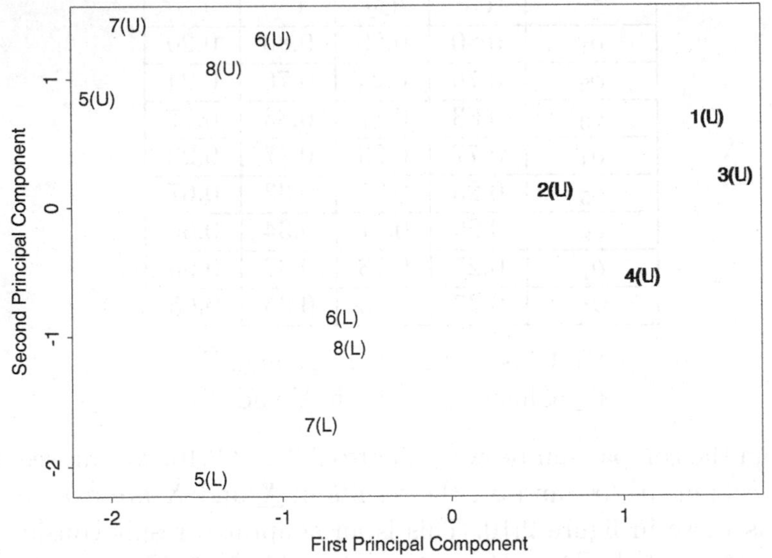

$1(L), 2(L), 3(L), 4(L), 5(L), 6(L), 7(L), 8(L)$: Eight Objects for $\underline{X}$
$1(U), 2(U), 3(U), 4(U), 5(U), 6(U), 7(U), 8(U)$: Eight Objects for $\overline{X}$

Fig. 2.10. Result of Fuzzy Clustering based WPCA for $\tilde{G}^*$ under Unique Clusters

Moreover, table 2.13 shows a comparison of the accumulated proportion of the first and the second principal components for $\tilde{G}$ and $\tilde{G}^*$ shown in equation (2.43). From this, we can see that we obtain a better result for $\tilde{G}^*$. That is, we can see that the WPCA considering uniqueness of clusters over minimum and maximum parts can obtain a better result when compared with the result of WPCA which does not consider the uniqueness of the clusters over the two parts.

Table 2.13. Comparison of Accumulated Proportions

Proportion for $\tilde{G}$	Proportion for $\tilde{G}^*$
0.902	0.913

The oil data shown in table 2.5 is used here [25]. The data is observed as interval-valued data. We divided the data into minimum values and maximum values and create the two data matrixes shown in table 2.6. Using the data, we create a super matrix G in equation (2.35). Applying G for a fuzzy clustering method named FANNY [27], we obtain the result shown in table 2.14. The number of clusters is assumed as 2. In table 2.14, each value shows the degree of belongingness of each oil to each cluster. $\underline{U}^*$ shows the result of degree of belongingness for $\underline{X}$, $\overline{U}^*$ is the result for $\overline{X}$, and they are shown in equation (2.36). C_1 shows cluster 1 and C_2 shows cluster 2. Notice that C_1 and C_2 in $\underline{U}^*$ are exactly the same as C_1 and C_2 in $\overline{U}^*$. Using the result in table 2.14, the weights shown in table 2.15 are obtained for the diagonal parts of $\underline{W}^*$ and $\overline{W}^*$ in equations (2.38) and (2.39), respectively. Using the data shown in table 2.5 and the weights in table 2.15, $\tilde{G}$ is calculated using equation (2.37).

Table 2.14. Fuzzy Clustering Results for Oil Data

Oils	$\underline{U}^*$		$\overline{U}^*$	
	C_1	C_2	C_1	C_2
Linseed Oil	0.65	0.35	0.95	0.05
Perilla Oil	0.93	0.07	0.94	0.06
Cottonseed Oil	0.12	0.88	0.23	0.77
Sesame Oil	0.16	0.84	0.26	0.74
Camellia Oil	0.06	0.94	0.05	0.95
Olive Oil	0.06	0.94	0.08	0.92
Beef Tallow	0.18	0.82	0.15	0.85
Hog Fat	0.13	0.87	0.08	0.92

C_1: Cluster 1, C_2: Cluster 2

Table 2.15. Weights of Oil for Minimum and Maximum Data

Oils	$\underline{W}^*$	$\overline{W}^*$
Linseed Oil	4.38	21.84
Perilla Oil	15.77	17.75
Cottonseed Oil	9.61	5.58
Sesame Oil	7.57	5.21
Camellia Oil	18.77	19.44
Olive Oil	16.99	14.31
Beef Tallow	6.86	7.75
Hog Fat	8.67	13.25

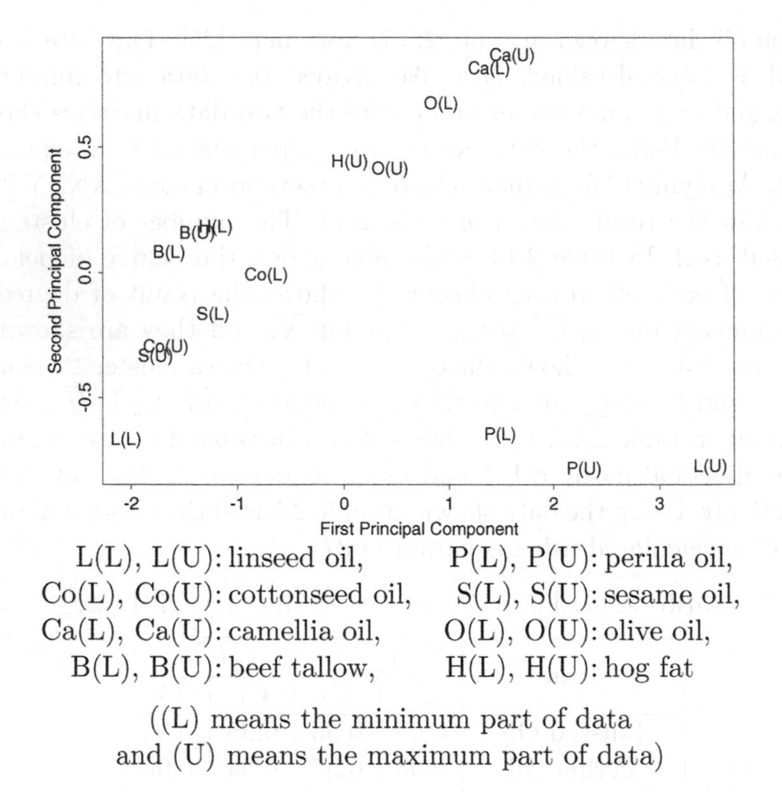

L(L), L(U): linseed oil, P(L), P(U): perilla oil,
Co(L), Co(U): cottonseed oil, S(L), S(U): sesame oil,
Ca(L), Ca(U): camellia oil, O(L), O(U): olive oil,
B(L), B(U): beef tallow, H(L), H(U): hog fat

((L) means the minimum part of data
and (U) means the maximum part of data)

Fig. 2.11. Result of WPCA under Unique Clusters for Oil Data

Figure 2.11 shows the result of WPCA based on fuzzy clustering for $\tilde{G}$ under the unique clusters over the minimum and maximum parts. In this figure, the abscissa shows the values of the first principal component shown in equation (2.41), and the ordinate is the values of the second principal component. As a reference, we show the result of conventional WPCA shown in equations (2.29) and (2.31) for the same data in figure 2.12. From the comparison between these two results shown in figures 2.11 and 2.12, we can see more details of the interval in figure 2.11. For example, in figure 2.12, we can see the similarity of "Ca" (camellia oil) and "O" (olive oil). And also we can see that "H" (hog fat) is in the middle position between "Ca" and "O", and "B" (beef tallow).

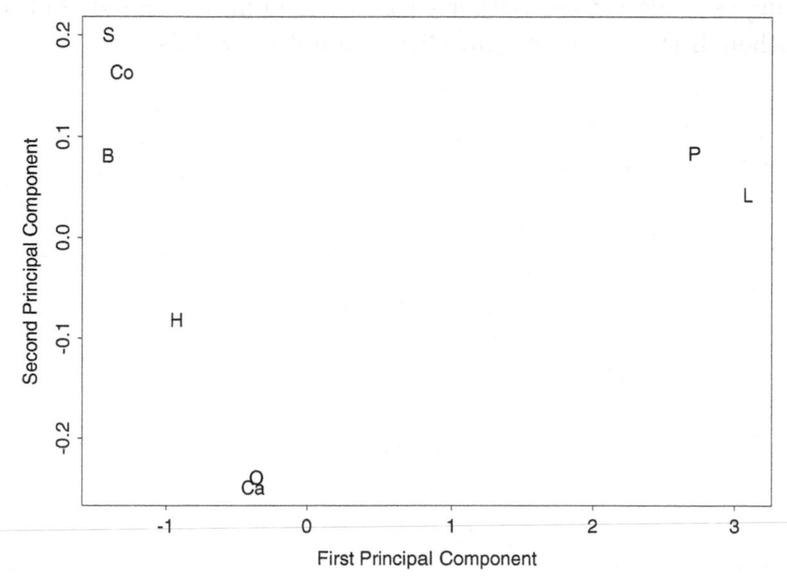

L: linseed oil, P: perilla oil, Co: cottonseed oil, S: sesame oil, Ca: camellia oil, O: olive oil, B: beef tallow, H: hog fat

Fig. 2.12. Result of WPCA for Oil Data

However, in figure 2.11, we can see the similarity of "Ca(U)", "Ca(L)", and "O(L)". That is, camellia oil is just similar to the minimum part of olive oil. While the maximum part of olive oil "O(U)" is similar to the maximum part of hog fat "H(U)". The minimum part of hog fat "H(L)" is similar to beef tallow "B". From this, we can see the reason for the result of "Ca", "O", and "H" in figure 2.12. So, we obtain more detailed information from the result shown in figure 2.11.

Moreover, from the value of the second principal component, we can see the similarity of "L" (linseed oil) and "P" (perilla oil) in figure 2.11. This is the same as the result in figure 2.12. In addition, we can see a remarkable difference between "L(L)" and "L(U)" for the first principal component in figure 2.11. This is reflected by the large difference between the minimum and the maximum values of the linseed oil with respect to the saponification value shown in table 2.5. We can see the very small value, 118, for the minimum value of linseed oil for the saponification, while minimum values of other oils for the saponification are large. That is, the result shown in figure 2.11 can show the difference of the minimum and maximum parts of the data, allowing us to detect the outlier such as the minimum value of linseed oil, although the result in figure 2.12 cannot show this.

3. Fuzzy Clustering based Regression Analysis

Regression analysis is widely used in many research areas including multivariate analysis. However, conventional regression analysis needs a linear structure for the observation space. In other words, if the given observation does not have a linear structure, then this method cannot capture the inherent factors of the data. In many cases it is not adequate. In order to solve this problem, regression methods which introduce a nonlinear structure to the regression model have been proposed [7], [10], [11], [19], [32]. Nonlinear regression models, generalized additive models, and weighted regression methods are typical examples of such methods.

Based on these ideas, we have proposed hybrid techniques of fuzzy clustering and regression methods [47], [55]. Fuzzy clustering itself has the ability to capture the nonlinear structure of data, however, we bring the data analysis ability into full play by a fusion of abilities which can capture the latent structure linearly and nonlinearly. We capture the linear structure while considering the classification structure of the observation. To crystallize the ideas, we focus on a fuzzy cluster loading model and a fuzzy clustering based regression model.

3.1 Fuzzy Cluster Loading Models

The essence of fuzzy clustering is to consider not only the belonging status to the assumed clusters, but also to consider how much the objects belong to the clusters. There is merit in representing the complex data situations which real data always have.

However, the interpretation of such a fuzzy clustering causes some confusion, because we sometimes think that objects which have a similar degree of belongingness can together form one more cluster. In order to obtain an interpretation of the fuzzy clustering result, we have proposed models of fuzzy cluster loading which can show the

Mika Sato-Ilic and Lakhmi C. Jain: *Innovations in Fuzzy Clustering*, StudFuzz **205**, 45–88 (2006)
www.springerlink.com

degree of relationship between the obtained clusters and variables of a given data. The following sections describe several fuzzy cluster loading models.

3.1.1 Fuzzy Cluster Loading for 2-Way Data

An observed 2-way data which is composed of n objects and p variables is denoted as $X = (x_{ia})$, $i = 1, \dots, n$, $a = 1, \dots, p$. We apply a fuzzy clustering method [5], [36] and obtain the result of the fuzzy clustering. That is, the degree of belongingness,

$$U = (u_{ik}), \ i = 1, \dots, n, \ k = 1, \dots, K.$$

We show the estimate obtained as $\hat{U} = (\hat{u}_{ik})$. Where K is the number of clusters and u_{ik} shows degree of belongingness of an object i to a cluster k. Usually u_{ik} is assumed to satisfy the condition shown in equation (1.1). Here we assume the condition (2.12) in order to avoid $u_{ik} = 0$. In order to obtain an interpretation of the fuzzy clustering result, we have proposed the following model [44], [47]:

$$\hat{u}_{ik} = \sum_{a=1}^{p} x_{ia}\eta_{ak} + \varepsilon_{ik}, \ i = 1, \dots, n, \ k = 1, \dots, K, \tag{3.1}$$

where, ε_{ik} is an error which is assumed to be a normally distributed random variable, with mean zero and variance σ^2, that is, $\varepsilon_{ik} \sim N(0, \sigma^2)$. η_{ak} shows the fuzzy degree and represents the amount of loading of a cluster k to a variable a. We call this fuzzy cluster loading. This parameter will show how each cluster can be explained by each variable.

The model (3.1) is rewritten as

$$\hat{\boldsymbol{u}}_k = X\boldsymbol{\eta}_k + \boldsymbol{\varepsilon}_k, \quad k = 1, \dots, K, \tag{3.2}$$

using

$$\hat{\boldsymbol{u}}_k = \begin{pmatrix} \hat{u}_{1k} \\ \vdots \\ \hat{u}_{nk} \end{pmatrix}, \ X = \begin{pmatrix} x_{11} & \cdots & x_{1p} \\ \vdots & \ddots & \vdots \\ x_{n1} & \cdots & x_{np} \end{pmatrix}, \ \boldsymbol{\eta}_k = \begin{pmatrix} \eta_{1k} \\ \vdots \\ \eta_{pk} \end{pmatrix}, \ \boldsymbol{\varepsilon}_k = \begin{pmatrix} \varepsilon_{1k} \\ \vdots \\ \varepsilon_{nk} \end{pmatrix},$$

where we assume

$$\varepsilon_{ik} \sim N(0, \sigma^2).$$

The estimate of least squares of $\boldsymbol{\eta}_k$ for equation (3.2) is obtained as follows:

$$\tilde{\boldsymbol{\eta}}_k = (X'X)^{-1}X'\hat{\boldsymbol{u}}_k, \tag{3.3}$$

by minimizing

$$\boldsymbol{\varepsilon}_k'\boldsymbol{\varepsilon}_k = \varepsilon_{1k}^2 + \cdots + \varepsilon_{nk}^2. \tag{3.4}$$

Using equation (3.2) and

$$\hat{U}_k = \begin{pmatrix} \hat{u}_{1k}^{-1} & \cdots & 0 \\ \vdots & \ddots & \vdots \\ 0 & \cdots & \hat{u}_{nk}^{-1} \end{pmatrix},$$

the model given in equation (3.1) can be rewritten again as:

$$\mathbf{1} = \hat{U}_k X \boldsymbol{\eta}_k + \boldsymbol{e}_k, \quad \boldsymbol{e}_k \equiv \hat{U}_k \boldsymbol{\varepsilon}_k, \quad k = 1, \ldots, K, \tag{3.5}$$

where,

$$\mathbf{1} = \begin{pmatrix} 1 \\ \vdots \\ 1 \end{pmatrix}, \quad \boldsymbol{e}_k = \begin{pmatrix} e_{1k} \\ \vdots \\ e_{nk} \end{pmatrix},$$

under the condition

$$e_{ik} \sim N(0, \sigma^2).$$

By minimizing

$$\boldsymbol{e}_k'\boldsymbol{e}_k = \boldsymbol{\varepsilon}_k'\hat{U}_k^2\boldsymbol{\varepsilon}_k, \tag{3.6}$$

we obtain an estimate of least squares of $\boldsymbol{\eta}_k$ for equation (3.5) as follows:

$$\tilde{\tilde{\boldsymbol{\eta}}}_k = (X'\hat{U}_k^2 X)^{-1}X'\hat{U}_k\mathbf{1}. \tag{3.7}$$

From equation (3.6), $\tilde{\tilde{\boldsymbol{\eta}}}_k$ is the estimate of weighted least squares of $\boldsymbol{\eta}_k$ in the weighted regression analysis. In conventional weighted regression analysis, it is common for variance and/or covariance of observations for the dependent variable to be used as the weights. Here, we use the feature of fuzzy clustering, that is, $\hat{u}_{ik}$ shows the weights of objects in a cluster k, therefore we can use $\hat{u}_{ik}$ directly for the weights of the weighted regression. Equation (3.6) can be rewritten as follows:

$$\boldsymbol{e}_k'\boldsymbol{e}_k = \boldsymbol{\varepsilon}_k'\hat{U}_k^2\boldsymbol{\varepsilon}_k = (\hat{u}_{1k}^{-1}\varepsilon_{1k})^2 + \cdots + (\hat{u}_{nk}^{-1}\varepsilon_{nk})^2. \tag{3.8}$$

From equation (3.8) and condition (2.12), we can see that if an object i belongs to the cluster k with a large degree, that is, when $\hat{u}_{ik}$ is large, then the error for the object i to the cluster k becomes smaller. In other words, if a and b are two degrees of belongingness of an object i to a cluster k and satisfy conditions $a < b, \quad a, b \in (0,1]$, then $(b^{-1}\varepsilon_{ik})^2 < (a^{-1}\varepsilon_{ik})^2$. So, $\tilde{\tilde{\eta}}_k$ is obtained considering not only the fitness of the model shown in equation (3.2) but also considering the classification structure of the data X.

3.1.2 Fuzzy Cluster Loading for 3-Way Data

Data which is composed of objects, variables, and situations (or times) is known as 3-way data. For 3-way data, a clustering of objects is not always coincident for all situations. This is because the values of variables vary with each situation. For 3-way data, a fuzzy clustering method has been proposed in [36] and relational fuzzy c-means for 3-way data has been discussed in chapter 1. As we described in chapter 1, the merit of this method is that the degree of belongingness of objects to clusters is determined independently for the situation. In contrast, the interpretation of the clusters obtained is difficult. This is because we could get only one clustering result through situations. In order to solve this problem, we propose fuzzy cluster loading models for 3-way data in [43], [46] which are able to detect the meaning of the clusters.

We have shown in [36] and chapter 1 that the clustering problem for 3-way data can be regarded as a multicriteria optimization problem. The merit of multicriteria fuzzy clustering is that Pareto efficient clustering could be obtained though several situations.

The data treated in multidimensional data analysis are mainly of the matrix type data or 2-way data. Typically, 2-way data consist of objects and attributes (or variables). For instance, the measurements of height and weight of children. In this case, height and weight are attributes and children are objects. On the other hand, typical 3-way data is composed of objects, attributes, and situations (or times). For instance, the measurements of height and weight of children at several ages on situations.

The 3-way data, which is observed by the values of p variables with respect to n objects for T times, is denoted by the following

$$X^{(t)} = (x_{ia}^{(t)}), \tag{3.9}$$

$$i = 1, \ldots, n, \quad a = 1, \ldots, p, \quad t = 1, \ldots, T.$$

Figure 3.1 shows the construction of the 3-way data.

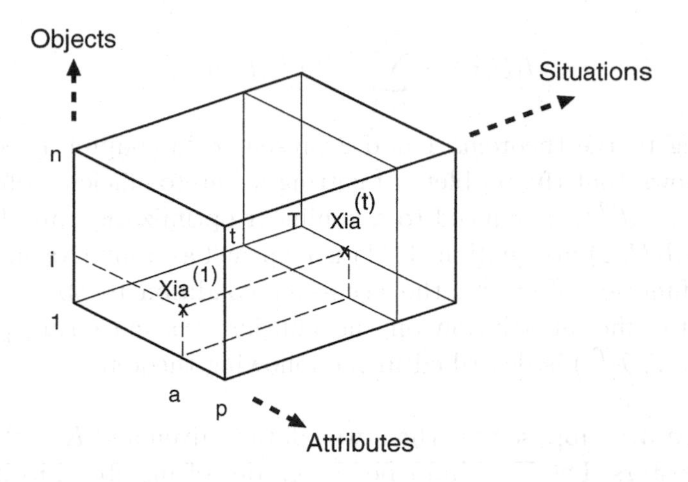

Fig. 3.1. 3-Way Data

The purpose of this clustering is to classify the n objects into K fuzzy clusters. The value of the membership functions of K fuzzy subsets, namely the degree of belongingness of each object to the cluster, is denoted by $U = (u_{ik})$, which satisfies condition (1.1). The centroid of the cluster at t-th time is denoted by

$$\boldsymbol{v}_k^{(t)} = (v_{ka}^{(t)}),$$

$$t = 1, \ldots, T, \quad k = 1, \ldots, K, \quad a = 1, \ldots, p.$$

The goodness of clustering in the t-th time is given by the sum of extended within-class dispersion [4],

$$J^{(t)}(U, \boldsymbol{v}^{(t)}) = \sum_{k=1}^{K} \sum_{i=1}^{n} (u_{ik})^m \sum_{a=1}^{p} (x_{ia}^{(t)} - v_{ka}^{(t)})^2. \tag{3.10}$$

The parameter m plays a role in determining the degree of fuzziness of the cluster. If there exists a solution $(U, \boldsymbol{v})$ which minimizes all $J^{(t)}$ $(t = 1, \ldots, T)$, then it is the best solution or dominant solution. Usually such a solution does not exist. Therefore this problem, a clustering for 3-way data, becomes a multicriteria optimization problem.

We assume that Φ is a set of feasible solutions $(U, \boldsymbol{v})$. We define a single clustering criterion by the weighted sum of $J^{(t)}$. That is, for $w^{(t)} > 0$,

$$J(U, \boldsymbol{v}) = \sum_{t=1}^{T} w^{(t)} J^{(t)}(U, \boldsymbol{v}^{(t)}). \tag{3.11}$$

According to the theorems 1 and 2 presented in chapter 1, equation (3.11) shows that the problem of getting a Pareto efficient solution of $(\Phi, J^{(1)}, \ldots, J^{(T)})$ is reduced to a nonlinear optimization problem.

Since $J(U, \boldsymbol{v})$ in equation (3.11) is regarded as a multivariable continuous function of $(U, \boldsymbol{v})$, the necessary condition to obtain the local Pareto efficient solution on the multicriteria clustering problem $(\Phi, J^{(1)}, \ldots, J^{(T)})$ is described in the following theorem:

Theorem 3.: Suppose that the 3-way data is given and K is the number of clusters. Let $U = (u_{ik})$ be the grades of membership in fuzzy clusters, and let $\boldsymbol{v}_k^{(t)} = (v_{ka}^{(t)})$ be a centroid of a cluster at t-th time. If $(U, \boldsymbol{v}) \in \Phi$ is at least a local Pareto efficient solution of multicriteria optimization problem $(\Phi, J^{(1)}, \ldots, J^{(T)})$ that is the minimization problem of equation (3.11), then $(U, \boldsymbol{v})$ satisfies the following equations (3.12) and (3.13). Let us assume

$$I_i = \{k \mid 1 \leq k \leq K;\ \sum_{t=1}^{T} d(\boldsymbol{x}_i^{(t)}, \boldsymbol{v}_k^{(t)}) = \sum_{t=1}^{T} \sum_{a=1}^{p} (x_{ia}^{(t)} - v_{ka}^{(t)})^2 = 0\},$$

$$\tilde{I}_i = \{1, 2, \ldots, K\} - I_i.$$

If $I_i = \phi$ and $\boldsymbol{v}_k^{(t)}$ is given, then

$$u_{ik} = \left[\sum_{l=1}^{K} \left\{ \frac{\sum_{t=1}^{T} w^{(t)} d(\boldsymbol{x}_i^{(t)}, \boldsymbol{v}_k^{(t)})}{\sum_{t=1}^{T} w^{(t)} d(\boldsymbol{x}_i^{(t)}, \boldsymbol{v}_l^{(t)})} \right\}^{\frac{1}{m-1}} \right]^{-1}, \tag{3.12}$$

$$i = 1, \ldots, n,\ k = 1, \ldots, K.$$

If $I_i \neq \phi$, then we define

$$u_{ik} = 0\ \forall k \in \tilde{I}_i,\ \sum_{k \in I_i} u_{ik} = 1.$$

For fixed u_{ik}, the centroid of each cluster is given by

$$v_{ka}^{(t)} = \frac{\sum_{i=1}^{n}(u_{ik})^m x_{ia}^{(t)}}{\sum_{i=1}^{n}(u_{ik})^m}, \tag{3.13}$$

$$t = 1, \ldots, T, \quad k = 1, \ldots, K, \quad a = 1, \ldots, p.$$

Note that the clusters are determined uniquely through the times for which $t = 1, \ldots, T$. However, the centroid of clusters $\boldsymbol{v}_k^{(t)}$ may differ with time. The purpose of the clustering (3.11) is to get common clusters through all of the times. That is, the grades u_{ik} are independent of the times.

The interpretation of the clusters created in equation (3.11) is difficult, because the degree of belongingness of objects to clusters, U, is determined independently for the situations. In order to solve this problem, we propose the following two models which have inherent fuzzy cluster loadings.

The models proposed are:

$$\hat{u}_{ik} = \sum_{a=1}^{p} x_{ia}^{(t)} \eta_{ak}^{(t)} + \varepsilon_{ik}^{(t)}, \quad i = 1, \ldots, n, \quad k = 1, \ldots, K, \quad t = 1, \ldots T. \tag{3.14}$$

$$\hat{u}_{ik} = \sum_{t=1}^{T} w^{(t)} \sum_{a=1}^{p} x_{ia}^{(t)} \eta_{ak} + \varepsilon_{ik}, \quad i = 1, \ldots, n, \quad k = 1, \ldots, K. \tag{3.15}$$

where, ε_{ik} and $\varepsilon_{ik}^{(t)}$ are errors and $\hat{u}_{ik}$ shows the obtained fuzzy clustering result as the degree of belongingness of an object i to a cluster k. We assume that these values are given as the clustering result in equation (3.11). $X^{(t)} = (x_{ia}^{(t)})$ is the given data matrix in equation (3.9). The solution of fuzzy clustering $\hat{U} = (\hat{u}_{ik})$ was obtained from this data. $\eta_{ak}^{(t)}$ in equation (3.14) shows the fuzzy degree which represents the amount of loading of a cluster k to a variable a at t-th time. This parameter shows how each cluster can be explained by each variable at each time. η_{ak} in equation (3.15) shows the fuzzy degree which represents the amount of loading of a cluster k to a variable a through all the times. The difference between the purpose of these two models is to capture the properties of the clusters that exist each time, or properties of the clusters over all of the times. $w^{(t)}$ shows the weights which are determined in equation (3.11).

$\eta_{ak}^{(t)}$ and η_{ak} are assumed to be satisfied by the following conditions:

$$\sum_{k=1}^{K} \eta_{ak} = \sum_{k=1}^{K} \eta_{ak}^{(t)} = 1, \quad a = 1, \ldots, p, \quad t = 1, \ldots, T, \tag{3.16}$$

$$\eta_{ak}, \ \eta_{ak}^{(t)} \in [0, 1], \quad a = 1, \ldots, p, \quad k = 1, \ldots, K, \quad t = 1, \ldots, T. \tag{3.17}$$

The purpose of equations (3.14) and (3.15) is to estimate $\eta_{ak}^{(t)}$ and η_{ak}, which minimize the following normalized sum of squared errors κ_1^2 and κ_2^2 under conditions (3.16) and (3.17).

$$\kappa_1^2 = \frac{\sum_{i=1}^{n}\sum_{k=1}^{K}(\hat{u}_{ik} - \sum_{a=1}^{p} x_{ia}^{(t)} \eta_{ak}^{(t)})^2}{\sum_{i=1}^{n}\sum_{k=1}^{K}(\hat{u}_{ik} - \bar{u})^2}, \quad t = 1, \ldots, T, \tag{3.18}$$

$$\kappa_2^2 = \frac{\sum_{i=1}^{n}\sum_{k=1}^{K}(\hat{u}_{ik} - \sum_{t=1}^{T} w^{(t)} \sum_{a=1}^{p} x_{ia}^{(t)} \eta_{ak})^2}{\sum_{i=1}^{n}\sum_{k=1}^{K}(\hat{u}_{ik} - \bar{u})^2}, \tag{3.19}$$

where,

$$\bar{u} = \frac{1}{nK} \sum_{i=1}^{n} \sum_{k=1}^{K} \hat{u}_{ik}.$$

In equation (3.14), we find the fuzzy cluster loadings $\eta_{ak}^{(t)}$ and α which minimize the following:

$$F_1 = \sum_{i=1}^{n}\sum_{k=1}^{K}(\hat{u}_{ik} - \alpha \sum_{a=1}^{p} x_{ia}^{(t)} \eta_{ak}^{(t)})^2, \quad t = 1, \ldots, T. \tag{3.20}$$

If $A_{ik}^{(t)} = \sum_{a=1}^{p} x_{ia}^{(t)} \eta_{ak}^{(t)}$, then equation (3.20) becomes

$$F_1 = \sum_{i=1}^{n}\sum_{k=1}^{K}(\hat{u}_{ik} - \alpha A_{ik}^{(t)})^2.$$

From $\dfrac{\partial F_1}{\partial \alpha} = 0$,

$$\alpha = \frac{\sum_{i=1}^{n}\sum_{k=1}^{K} \hat{u}_{ik} A_{ik}^{(t)}}{\sum_{i=1}^{n}\sum_{k=1}^{K} (A_{ik}^{(t)})^2}. \tag{3.21}$$

The descent vector is determined as follows:

$$\frac{\partial F_1}{\partial \eta_{bc}^{(t)}} = -2\alpha \sum_{i=1}^{n} (\hat{u}_{ic} - \alpha A_{ic}^{(t)}) x_{ib}^{(t)}, \quad b = 1, \ldots, p, \quad c = 1, \ldots, K. \tag{3.22}$$

The following method of descent is used to find the solutions $\eta_{ak}^{(t)}$ and α.

(Step 1) Fix K, $(2 \leq K < n)$.

(Step 2) Initialize $H^{(t)}(0) = (\eta_{ak}^{(t)}(0))$, $(a = 1, \ldots, p,\ k = 1, \ldots, K,\ t = 1, \ldots, T)$. And set the step number $q = 0$.

(Step 3) Set $q = q + 1$. Calculate $\alpha(q)$ by using equation (3.21).

(Step 4) Calculate the value of $\dfrac{\partial F_1}{\partial \eta_{ak}^{(t)}(q-1)}$ by using equation (3.22) and $\alpha(q)$ obtained in Step 3. Find the optimal solution with respect to the direction of the descent vector by using a one dimensional direct search. That is, update $\eta_{ak}^{(t)}$ by using the next expression,

$$\eta_{ak}^{(t)}(q) = \eta_{ak}^{(t)}(q-1) - \lambda \left(\frac{\partial F_1}{\partial \eta_{ak}^{(t)}(q-1)} \right),$$

where $\lambda > 0$ is the step size.

(Step 5) Calculate the value of $\| H^{(t)}(q) - H^{(t)}(q-1) \|$ by using $\alpha(q)$ and $\eta_{ak}^{(t)}(q)$ obtained in Step 3 and Step 4 respectively. Here $\| \cdot \|$ shows the norm of matrix.

(Step 6) If $\| H^{(t)}(q) - H^{(t)}(q-1) \| < \varepsilon$, then stop, or otherwise go to Step 3.

In equation (3.15), we should find the fuzzy cluster loadings η_{ak} and δ which minimize the following:

$$F_2 = \sum_{i=1}^{n}\sum_{k=1}^{K} (\hat{u}_{ik} - \delta \sum_{t=1}^{T} w^{(t)} \sum_{a=1}^{p} x_{ia}^{(t)} \eta_{ak})^2. \tag{3.23}$$

If $B_{ik} = \sum_{t=1}^{T} w^{(t)} \sum_{a=1}^{p} x_{ia}^{(t)} \eta_{ak}$, then equation (3.23) is

$$F_2 = \sum_{i=1}^{n} \sum_{k=1}^{K} (\hat{u}_{ik} - \delta B_{ik})^2.$$

From $\frac{\partial F_2}{\partial \delta} = 0$,

$$\delta = \frac{\sum_{i=1}^{n} \sum_{k=1}^{K} \hat{u}_{ik} B_{ik}}{\sum_{i=1}^{n} \sum_{k=1}^{K} (B_{ik})^2}. \tag{3.24}$$

The descent vector is determined as follows:

$$\frac{\partial F_2}{\partial \eta_{bc}} = -2\delta \sum_{i=1}^{n} (\hat{u}_{ic} - \delta B_{ic}) \sum_{t=1}^{T} w^{(t)} x_{ib}^{(t)}, \quad b = 1, \ldots, p, \quad c = 1, \ldots, K. \tag{3.25}$$

The following method of descent is used to find the solutions η_{ak} and δ.

(Step 1) Fix K, $(2 \leq K < n)$, and $w^{(t)}$, $(t = 1, \ldots, T)$.

(Step 2) Initialize $H(0) = (\eta_{ak}(0))$, $(a = 1, \ldots, p,\ k = 1, \ldots, K)$. Set the step number $q = 0$.

(Step 3) Set $q = q + 1$. Calculate $\delta(q)$ by using equation (3.24).

(Step 4) Calculate the value of $\frac{\partial F_2}{\partial \eta_{ak}(q-1)}$ by using equation (3.25) and $\delta(q)$ obtained in Step 3. Find the optimal solution with respect to the direction of the descent vector by using a one dimensional direct search. That is, update η_{ak} by using the expression,

$$\eta_{ak}(q) = \eta_{ak}(q-1) - \lambda \left(\frac{\partial F_2}{\partial \eta_{ak}(q-1)} \right),$$

where $\lambda > 0$ is step size.

(Step 5) Calculate the value of $\|H(q) - H(q-1)\|$ by using $\delta(q)$ and $\eta_{ak}(q)$ obtained in Step 3 and Step 4 respectively, where $\|\cdot\|$ shows the norm of matrix.

(Step 6) If $\|H(q) - H(q-1)\| < \varepsilon$, stop, or otherwise go to Step 3.

3.1.3 Fuzzy Cluster Loading for Asymmetric 2-Way Data

Recently, clustering techniques based on asymmetric similarity (proximity) data have generated tremendous interest among a number of researchers. For example, two-way communication in human relationships, information flow, or data of two-way communication system are typical examples of such asymmetric similarity data.

In this section, we describe fuzzy cluster loading models for asymmetric similarity data [49]. These models use the technique of the dynamic clustering model for 3-way data [37]. We decompose the upper triangle part and the lower triangle part of the asymmetric similarity data matrix. This makes two symmetric matrixes corresponding to the upper and lower triangle parts. Here, the upper part means one way communication from the origin to the destination, and the lower part means one way communication from the destination to the origin. Using these matrixes and the average function family, we reconstruct the data as a super matrix. From this matrix and by using the dynamic fuzzy clustering model, we can get the solution as the degree of belongingness of objects to clusters over the two matrixes (that is over the upper and lower parts). The merit of this idea is to get the degree of belongingness under the unique coordinate of the invariant clusters. That is, the clusters of the upper part of the asymmetric similarity data are exactly the same as the clusters of the lower part. We can then treat the solutions of these two parts as comparable.

The interpretations of the obtained clusters for the upper part and the lower part may differ with each other. So, we use the idea of fuzzy cluster loading which captures the quantities of relation between variables and clusters directly. From the model which shows the fuzzy cluster loading, we can estimate the fuzzy cluster loading with respect to the same variables and clusters. Therefore, we can compare the two results of fuzzy cluster loading for both total interpretation of the clusters, and the interpretation of the clusters of either the upper part or the lower part of the similarity data. If the given asymmetric data is two-way communication data, then this result shows that we can capture the features of the clusters obtained for two-way communication data and the feature of the one way communication part of asymmetric similarity data. These features are comparable with each other due to the fuzzy cluster loading model. This model can get the loadings with respect to the same clusters and variables for the result of degree of belongingness of two-way and one-way communications.

We first describe the dynamic additive fuzzy clustering model. The data is observed by the values of similarity with respect to n objects for T times. The similarity matrix of t-th time is shown by $S^{(t)} = (s_{ij}^{(t)})$, $i, j = 1, \ldots, n$. Then a $Tn \times Tn$ matrix $\tilde{S}$ is denoted as follows:

$$\tilde{S} = \begin{bmatrix} S^{(1)} & S^{(12)} & S^{(13)} & \ldots & S^{(1T)} \\ S^{(21)} & S^{(2)} & S^{(23)} & \ldots & S^{(2T)} \\ \vdots & \vdots & \vdots & & \vdots \\ S^{(T1)} & S^{(T2)} & S^{(T3)} & \ldots & S^{(T)} \end{bmatrix}. \tag{3.26}$$

In equation (3.26), the diagonal matrix is the $n \times n$ matrix $S^{(t)}$. $S^{(rt)}$ is a $n \times n$ matrix and the element is defined as $s_{ij}^{(rt)} \equiv m(s_{ij}^{(r)}, s_{ij}^{(t)}),\ r, t = 1, \ldots, T$, where $s_{ij}^{(t)}$ is the (i, j)-th element of the matrix $S^{(t)}$. $m(x, y)$ is an average function described in chapter 1. The examples of the average function are shown in table 1.1. For the element $\tilde{s}_{ij}$ of $\tilde{S}$, we have proposed the dynamic additive fuzzy clustering model [37] as:

$$\tilde{s}_{ij} = \sum_{k=1}^{K} \rho(u_{i^{(t)}k}, u_{j^{(t)}k}), \quad i, j = 1, \ldots, Tn, \tag{3.27}$$

$$i^{(t)} \equiv i - n(t-1), \ (n(t-1) + 1 \leq i \leq tn, \ t = 1, \ldots, T),$$

where K is a number of clusters and $u_{i^{(t)}k}$ shows the degree of belongingness of an object i to a cluster k at time t. Here $i^{(t)} = 1, \ldots, n, \ t = 1, \ldots, T$. $u_{i^{(t)}k}$ satisfies

$$u_{i^{(t)}k} \geq 0, \ \sum_{k=1}^{K} u_{i^{(t)}k} = 1, \ \forall i, t.$$

$\rho(u_{i^{(t)}k}, u_{j^{(t)}k})$ is assumed to be sharing a common property of objects i and j to a cluster k at t-th time and ρ is a symmetric aggregation function which satisfies the following conditions where $x, y, z, w \in [0, 1]$:

$$0 \leq \rho(x, y) \leq 1, \ \rho(x, 0) = 0, \ \rho(x, 1) = x. \ \text{(Boundary Condition)}$$

$$\rho(x, y) \leq \rho(z, w) \ \text{ whenever } \ x \leq z, \ y \leq w. \ \text{(Monotonicity)}$$

$$\rho(x, y) = \rho(y, x). \ \text{(Symmetry)}$$

In particular, t-norm [33], [59] is well-known as a concrete example which satisfies the above conditions. If the observed similarity is asymmetric, then the model (3.27) cannot be used. In order to apply the asymmetric data to the model (3.27), we have to reconstruct the super matrix (3.26). Suppose

$$S^{(tU)} = (s_{ij}^{(tU)}),\ s_{ij}^{(tU)} = \begin{cases} s_{ij}^{(t)}, i \le j \\ s_{ji}^{(t)}, i > j \end{cases},$$

$$S^{(tL)} = (s_{ij}^{(tL)}),\ s_{ij}^{(tL)} = \begin{cases} s_{ij}^{(t)}, i \ge j \\ s_{ji}^{(t)}, i < j \end{cases}. \tag{3.28}$$

We then assume the super matrix as:

$$\tilde{\tilde{S}} = \begin{bmatrix} S^{(1U)} & S^{(1U)(1L)} & \dots & S^{(1U)(TL)} \\ S^{(1L)(1U)} & S^{(1L)} & \dots & S^{(1L)(TL)} \\ S^{(2U)(1U)} & S^{(2U)(1L)} & \dots & S^{(2U)(TL)} \\ \vdots & \vdots & \vdots & \vdots \\ S^{(TL)(1U)} & S^{(TL)(1L)} & \dots & S^{(TL)} \end{bmatrix},$$

where

$$S^{(rU)(tL)} = (s_{ij}^{(rU)(tL)}),\ s_{ij}^{(rU)(tL)} \equiv m(s_{ij}^{(rU)}, s_{ij}^{(tL)}). \tag{3.29}$$

Equation (3.27) is rewritten as

$$\tilde{\tilde{s}}_{ij} = \sum_{k=1}^{K} \rho(u_{i^{(c)}d_k}, u_{j^{(c)}d_k}),\ \ i, j = 1, \dots, 2Tn, \tag{3.30}$$

$$i^{(c)} \equiv i - n(c-1),\ (n(c-1)+1 \le i \le cn,\ c = 1, \dots, 2T).$$

$\tilde{\tilde{s}}_{ij}$ is (i, j)-th element of $\tilde{\tilde{S}}$ and $u_{i^{(c)}d_k}$ is a degree of belongingness of an object i to a cluster k, and c and d are suffixes for representing time and the direction of the asymmetric proximity data respectively. The relationship is shown as follows:

$$d = \begin{cases} c \equiv 0 \pmod 2,\ d = L \text{ and } t = \frac{c}{2} \\ c \equiv 1 \pmod 2,\ d = U \text{ and } t = \frac{c+1}{2} \end{cases},$$

where $t = 1, \dots, T$. L and U show the lower and upper triangular matrixes, respectively, that is, an asymmetric relationship in the same time. t shows t-th time, and $c \equiv 0 \pmod 2$ means that c is congruent to 0 modulo 2.

Suppose $X = (x_{ia}),\ i = 1, \ldots, n,\quad a = 1, \ldots, p$ be some observed 2-way data and we know that there is an asymmetric relation among n objects. In order to create an asymmetric similarity matrix $S = (s_{ij}),\ \ s_{ij} \neq s_{ji},\ \ (i \neq j,\ i, j = 1, \ldots, n)$, we use the following asymmetric aggregation function [39]:
Denoting the asymmetric aggregation function by $s(x, y)$, it is defined as follows:
Suppose that $f(x)$ is a generating function of t-norm [33], [59] and $\phi(x)$ is a continuous monotone decreasing function satisfying $\phi : [0, 1] \rightarrow [1, \infty],\ \ \phi(1) = 1$. We define the asymmetric aggregation operator $s(x, y)$ as:

$$s(x, y) = f^{[-1]}(f(x) + \phi(x) f(y)). \tag{3.31}$$

Note that $s(x, y)$ satisfies boundary and monotonicity conditions of t-norm and asymmetric condition $s(x, y) \neq s(y, x)$. Using this asymmetric aggregation function s, we obtain $s_{ij} = s(\boldsymbol{x}_i, \boldsymbol{x}_j)(\neq s_{ji})$, where $\boldsymbol{x}_i = (x_{i1}, \ldots, x_{ip})$.

We create the upper triangular matrix $S^{(U)}$, the lower triangular matrix $S^{(L)}$, and $S^{(U)(L)}, S^{(L)(U)}$ from S according to equations (3.28) and (3.29) when $t, r = 1$. That is, we make the following super matrix:

$$\tilde{\tilde{S}} = \begin{bmatrix} S^{(U)} & S^{(U)(L)} \\ S^{(L)(U)} & S^{(L)} \end{bmatrix}. \tag{3.32}$$

Using equation (3.30), we obtain the degree of belongingness of objects to clusters:

$$U = \begin{pmatrix} U^{(U)} \\ U^{(L)} \end{pmatrix}, \tag{3.33}$$

where the $n \times K$ matrix $U^{(U)} = (u_{ik}^{(U)})$ shows the result of degree of belongingness of the objects to K clusters from $S^{(U)}$. The $n \times K$ matrix $U^{(L)} = (u_{ik}^{(L)})$ shows the result of $S^{(L)}$. The merit of this clustering is that we can compare the two results $U^{(U)}$ and $U^{(L)}$, due to the same K clusters. However, the interpretation of the K clusters may be different for $S^{(U)}$ and $S^{(L)}$. By using model (3.1) we compare the interpretation of the clusters under $S^{(U)}$ and $S^{(L)}$. From model (3.1), we can write the following models for capturing the properties of the clusters of $S^{(U)}$ and $S^{(L)}$.

$$\hat{u}_{ik}^{(U)} = \sum_{a=1}^{p} x_{ia}\eta_{ak}^{(U)} + \varepsilon_{ik}, \quad \hat{u}_{ik}^{(L)} = \sum_{a=1}^{p} x_{ia}\eta_{ak}^{(L)} + \varepsilon_{ik}, \tag{3.34}$$

$$i = 1, \ldots, n, \ k = 1, \ldots, K.$$

Here $\eta_{ak}^{(U)}$ shows fuzzy cluster loading of a variable a to a cluster k for the upper triangle matrix $S^{(U)}$ and $\eta_{ak}^{(L)}$ shows one for the lower triangle matrix $S^{(L)}$. $\hat{u}_{ik}^{(U)}$ and $\hat{u}_{ik}^{(L)}$ are estimates obtained in equation (3.33). Note that the cluster k and variable a of $\eta_{ak}^{(U)}$ are the same as the cluster k and variable a of $\eta_{ak}^{(L)}$. So, we can compare the fuzzy cluster loadings $\eta_{ak}^{(U)}$ and $\eta_{ak}^{(L)}$.

3.1.4 Fuzzy Cluster Loading for Asymmetric 3-Way Data

In order to obtain the dynamic interpretation for times and the direction of asymmetry for 3-way asymmetric data, we extend the fuzzy cluster loading model for the 3-way asymmetric data. These fuzzy cluster loadings have the following merits that (1) although the obtained clusters are the same over the times or the direction of asymmetry, the interpretation of the clusters (fuzzy cluster loading) can be obtained so as to be changeable according to the times or the direction of asymmetry. (2) the interpretations of the clusters (fuzzy cluster loadings) at each time point and each direction of asymmetry are mathematically comparable under the invariant clusters (the unique coordinate) over the times or the direction of asymmetry. (3) we can compare the result of fuzzy cluster loadings at each time and each direction, and the result of the dynamic fuzzy clustering under the invariant clusters.

If we obtain the result of fuzzy clustering for 3-way asymmetric similarity data, $\hat{u}_{i^{(c)}dk}$, then using model (3.1) and 3-way data for the same objects and times, we propose the following model:

$$\hat{u}_{i^{(c)}dk} = \sum_{a=1}^{p} x_{i^{(c)}a}\eta_{adk}^{(c)} + \varepsilon_{i^{(c)}dk}, \quad \forall i^{(c)}, k, \tag{3.35}$$

$$i, j = 1, \ldots, 2Tn, \ i^{(c)} \equiv i - n(c-1), \ (n(c-1)+1 \le i \le cn, \ c = 1, \ldots, 2T),$$

$$d = \begin{cases} c \equiv 0 \pmod 2, d = L \text{ and } t = \frac{c}{2} \\ c \equiv 1 \pmod 2, d = U \text{ and } t = \frac{c+1}{2} \end{cases},$$

where $\eta_{a^d k}^{(c)}$ is a fuzzy cluster loading which shows the fuzzy degree representing the amount of loading of a cluster k to a variable a at d direction (upper part or lower parts). $\varepsilon_{i^{(c)}d_k}$ is an error. Equation (3.35) is a generalized model of equation (3.34). Equation (3.35) is rewritten as:

$$\mathbf{1} = \hat{U}_{dk}^{(c)} X^{(c)} \boldsymbol{\eta}_{dk}^{(c)} + \boldsymbol{e}_{dk}^{(c)}, \quad \boldsymbol{e}_{dk}^{(c)} \equiv \hat{U}_{dk}^{(c)} \boldsymbol{\varepsilon}_{dk}^{(c)}, \tag{3.36}$$

using

$$\hat{U}_{dk}^{(c)} = \begin{pmatrix} \hat{u}_{1^{(c)}d_k}^{-1} & 0 & \cdots & 0 \\ 0 & \hat{u}_{2^{(c)}d_k}^{-1} & \cdots & \vdots \\ \vdots & \vdots & \ddots & \vdots \\ 0 & \cdots & \cdots & \hat{u}_{n^{(c)}d_k}^{-1} \end{pmatrix}, \quad X^{(c)} = \begin{pmatrix} x_{1^{(c)}1} & \cdots & x_{1^{(c)}p} \\ x_{2^{(c)}1} & \cdots & x_{2^{(c)}p} \\ \vdots & \vdots & \vdots \\ x_{n^{(c)}1} & \cdots & x_{n^{(c)}p} \end{pmatrix},$$

$$\mathbf{1} = \begin{pmatrix} 1 \\ 1 \\ \vdots \\ 1 \end{pmatrix}, \quad \boldsymbol{\eta}_{dk}^{(c)} = \begin{pmatrix} \eta_{1^d k}^{(c)} \\ \eta_{2^d k}^{(c)} \\ \vdots \\ \eta_{p^d k}^{(c)} \end{pmatrix}.$$

Where $X^{(c)} (c = 1, \ldots, 2T)$ is a data matrix of n objects with respect to p variables at t-th time. Here $t = \frac{c}{2}$ $(c \equiv 0 \pmod 2)$, $t = \frac{c+1}{2}$ $(c \equiv 1 \pmod 2)$. The objects of $X^{(t)}$ are the same as the objects of $S^{(t)}$, $t = 1, \ldots, T$.

From equation (3.36), we obtain the estimate of $\boldsymbol{\eta}_{dk}^{(c)}$ as:

$$\tilde{\boldsymbol{\eta}}_{dk}^{(c)} = ((X^{(c)})' (\hat{U}_{dk}^{(c)})^2 X^{(c)})^{-1} (X^{(c)})' \hat{U}_{dk}^{(c)} \mathbf{1}. \tag{3.37}$$

3.2 Fuzzy Clustering based Weighted Regression Model

We describe another model for weighted regression using a fuzzy clustering result obtained as a classification of the data consisting of independent variables [55]. This model is closely related to geographically weighted regression analysis [7] and the fuzzy c-regression model [23]. This model is for the hybrid techniques of clustering and regression. The merit of geographically weighted regression analysis is to consider the spatial feature over several areas using the weights estimated by the kernel function for a regression model and consider the difference among the features of the areas. The fuzzy clustering based regression model retains this important feature and highlights that the classification of objects into the areas is obtained as the result of data of independent variables.

The essence of fuzzy clustering is to consider not only the belonging status of objects to clusters but also to consider how much the objects belong to the clusters. That is, the result of the clustering is represented as the degree of belongingness of objects to clusters. This is unlike the conventional clustering result, which only shows whether the object does or does not belong to a cluster. Therefore, conventional clustering obtains the exclusive clusters, but fuzzy clustering obtains the overlapping clusters.

The support for the distribution of the degree of objects to the overlapping clusters is all objects, so we can treat the degree in a similar way as the estimate using the kernel function in geographically weighted regression analysis, by regarding each obtained cluster as each area. In other words, we can use the degree of belongingness of objects to clusters (areas) as the weights in geographically weighted regression analysis, in order to obtain the estimate of the weights for each classified area.

As in conventional research for geographically weighted regression analysis, an extension for the functional data has been proposed [66]. Also, as a method of data analysis using the idea of the estimate of the weighted regression analysis and the result of fuzzy clustering, a model of fuzzy cluster loadings has been described previously. The purpose of fuzzy cluster loading is to get the interpretation of the obtained fuzzy cluster. It is known that the estimate of the fuzzy cluster loading is reduced to the regression coefficients for the weighted regression model. The weights are treated as the degree of belongingness of the objects to the clusters which is similar to the proposed regression model in this section. However, the weights in the geographically weighted regression model are replaced by the degree of belongingness of objects to clusters directly. Otherwise in the case of the model of the fuzzy cluster loadings, the process of the estimation of the fuzzy cluster loadings, the weights are indirectly reduced to be treated as the degree of belongingness of objects to clusters.

Based on this idea we try to bring data analysis into play by introducing pre-information which is represented as weights of an observed data to the regression model, a possibilistic regression model [62] and the fuzzy c-regression model [23] are among several examples of research that introduced the concept of fuzziness to the regression model. In particular, the fuzzy c-regression model estimates weights by using fuzzy clustering. In the case of the geographically weighted regression model, the weights are estimated by using the kernel method. Both of the methods use the weighted least squares method in order to estimate the regression coefficients. Using conventional weighted least squares method means that we use the same weights for the data with respect to the independent variables, and also the data with respect to a dependent variable. This is not always adaptable when using a classification structure for the weights. This is because the data with respect to the dependent variable, usually have a different classification structure from the data with respect to the independent variables.

So, we proposed a weighted regression model based on fuzzy clustering in which we use only the classification structure of the data with respect to independent variables and use the weights only for the data with respect to independent variables.

The model is defined as:

$$\boldsymbol{y} = \hat{U}_k^{-1} X \boldsymbol{\beta}_k + \hat{\boldsymbol{e}}_k, \tag{3.38}$$

where

$$\hat{U}_k^{-1} = \begin{pmatrix} \hat{u}_{1k} & \cdots & 0 \\ \vdots & \ddots & \vdots \\ 0 & \cdots & \hat{u}_{nk} \end{pmatrix}, \quad \hat{\boldsymbol{e}}_k = \begin{pmatrix} \hat{e}_{1k} \\ \vdots \\ \hat{e}_{nk} \end{pmatrix}, \quad \hat{e}_{ik} \sim N(0, \sigma^2).$$

The estimate of least squares of $\boldsymbol{\beta}_k$ in equation (3.38) is obtained as:

$$\tilde{\boldsymbol{\beta}}_k = (X'(\hat{U}_k^{-1})^2 X)^{-1} X' \hat{U}_k^{-1} \boldsymbol{y}. \tag{3.39}$$

Related to equation (3.39), the fuzzy c-regression model [23] has been proposed. This model essentially obtains the following estimate of weighted least squares:

$$\tilde{\tilde{\boldsymbol{\beta}}}_k = (X'(\hat{U}_k^{-1})^m X)^{-1} X' (\hat{U}_k^{-1})^m \boldsymbol{y}, \tag{3.40}$$

which minimizes

$$S = \sum_{k=1}^{K} S_k,$$

$$S_k \equiv (\boldsymbol{y} - X\boldsymbol{\beta}_k)' (\hat{U}_k^{-1})^m (\boldsymbol{y} - X\boldsymbol{\beta}_k) \equiv \boldsymbol{\delta}_k' \boldsymbol{\delta}_k, \tag{3.41}$$

$$\boldsymbol{\delta}_k \equiv (\hat{U}_k^{-1})^{\frac{m}{2}}(\boldsymbol{y} - X\boldsymbol{\beta}_k) = \begin{pmatrix} \delta_{1k} \\ \vdots \\ \delta_{nk} \end{pmatrix}, \quad m > 1,$$

where

$$(\hat{U}_k^{-1})^m = \begin{pmatrix} \hat{u}_{1k}^m & \cdots & 0 \\ \vdots & \ddots & \vdots \\ 0 & \cdots & \hat{u}_{nk}^m \end{pmatrix}, \quad \delta_{ik} \sim N(0, \sigma^2).$$

Notice that equation (3.40) is different from equation (3.39), even if we put $m = 2$ in equation (3.40). The structure of equation (3.40) is essentially the same as the estimate of conventional weighted regression analysis. That is, in equation (3.41), we assume that weights of objects in the dependent variable are the same as the weights of objects in the independent variables. However, in equation (3.38), we consider the difference of data structure between dependent variable and independent variables. So, we multiplied the weights only to X in equation (3.38), since $\hat{U}_k^{-1}$ is obtained as a classification result of X.

The geographically weighted regression model [7] has been proposed. The estimate of regression coefficient to an area k is obtained as:

$$\hat{\boldsymbol{\beta}}_k = (X'A_kX)^{-1}X'A_k\boldsymbol{y}, \tag{3.42}$$

where

$$A_k = \begin{pmatrix} \alpha_{1k} & \cdots & 0 \\ \vdots & \ddots & \vdots \\ 0 & \cdots & \alpha_{nk} \end{pmatrix}, \quad k = 1, \ldots, K.$$

K gives the number of areas and α_{ik} gives the weight of an object i to an area k estimated by using kernel estimates. That is, the weights A_k are defined as:

$$\alpha_{ik} = \phi(d_{ik}), \quad i = 1, \ldots, n, \; k = 1, \ldots, K,$$

by using the kernel function ϕ which satisfies the following conditions:

$$\phi(0) = 1,$$

$$\lim_{d \to \infty} \{\phi(d)\} = 0,$$

$$\phi \text{ is a monotone decreasing function,}$$

where d_{ik} shows Euclidean distance between i-th object and a centroid of the k-th area.

The following are examples of ϕ:

$$\alpha_{ik} = \exp\left(\frac{-d_{ik}}{r}\right), \tag{3.43}$$

$$\alpha_{ik} = \exp\left(\frac{-d_{ik}^2}{2r^2}\right),$$

where r shows the control parameter influenced by the range of each area.

Equation (3.42) is obtained as minimizing

$$\hat{S}_k \equiv (\boldsymbol{y} - X\boldsymbol{\beta}_k)' A_k (\boldsymbol{y} - X\boldsymbol{\beta}_k) \equiv \hat{\boldsymbol{\delta}}_k' \hat{\boldsymbol{\delta}}_k, \tag{3.44}$$

$$\hat{\boldsymbol{\delta}}_k \equiv A_k^{\frac{1}{2}} (\boldsymbol{y} - X\boldsymbol{\beta}_k) = \begin{pmatrix} \hat{\delta}_{1k} \\ \vdots \\ \hat{\delta}_{nk} \end{pmatrix},$$

where

$$A_k^{\frac{1}{2}} = \begin{pmatrix} \sqrt{\alpha_{1k}} & \cdots & 0 \\ \vdots & \ddots & \vdots \\ 0 & \cdots & \sqrt{\alpha_{nk}} \end{pmatrix}, \quad \hat{\delta}_{ik} \sim N(0, \sigma^2).$$

Equation (3.42) is similar to equation (3.40). The difference is only the weights. Equation (3.42) is used for the weights estimated by the kernel method, otherwise, equation (3.40) is obtained by using the weights estimated by fuzzy clustering.

3.3 Numerical Examples for Regression Models based on Fuzzy Clustering

3.3.1 Numerical Examples for Fuzzy Cluster Loading for 2-Way Data

We first use artificially created data in order to investigate the validity of model (3.1). The data is shown in figure 3.2. In this figure, eight lines show eight objects with respect to seven variables $v_1, \ldots, v_7$. The abscissa shows each of the variables and ordinate shows the values for each of the variables. The solid lines show the objects o_1, o_2, o_3, and o_4 and the dotted lines show the objects o_5, o_6, o_7, and o_8.

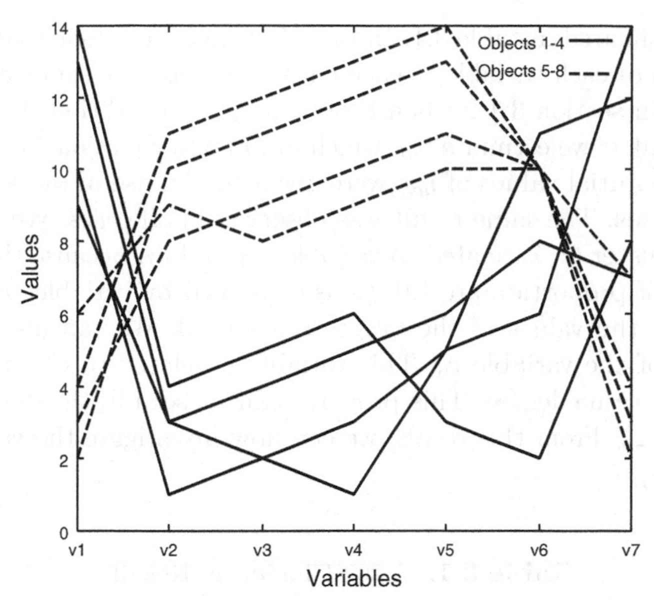

Fig. 3.2. Artificial Data

From this figure, we can see the similarity between objects o_1, o_2, o_3 and o_4, also that between objects o_5, o_6, o_7 and o_8. The feature of this data is that the first group which consists of objects o_1, o_2, o_3, and o_4 has the property that the values of variables v_1 and v_7 are large and values of variables v_2 - v_5 are small. The second group has the opposite property, excluding the value of variable v_6. In variable v_6, we can see that the two groups are mixed.

The result of the fuzzy c-means method using $m = 2.0$ (given in equation (1.3)) is shown in table 3.1. In this table, each value shows degree of belongingness to each cluster, cluster C_1 and cluster C_2. o_1 - o_8 show the eight objects. From this result we can see that objects o_1 - o_4 for the most part belong to cluster C_1 and objects o_5 - o_8 belong to cluster C_2. According to the feature of this data, created with the intention of forming two clusters, this result is adaptable.

Usually we do not know the property of the data structure shown in figure 3.2, so after clustering we get the result shown in table 3.1. The problem in this case is how to find the properties of clusters C_1 and C_2. In order to solve this problem, we use the model shown in equation (3.1). Table 3.2 shows the result of model (3.1) using the fuzzy clustering result shown in table 3.1 and the data shown in figure 3.2. In table 3.2, $v_1, \ldots, v_7$ show seven variables and C_1, C_2 are the two clusters indicated in table 3.1. In table 3.2, each value shows degree of proportion of each variable to each cluster. We use the same algorithm discussed in section 3.1.2 when $t = 1$ under the conditions (3.16) and (3.17). That is we estimate η_{ak} which minimizes equation (3.18) when $t = 1$. The initial values of η_{ak} were given by the use of random numbers 20 times. The same result was observed in all cases. We can now see that cluster C_1 is related to variables v_1 and v_7, because the values of degree of proportion are 1.0. C_2 is explained by variables v_2, v_3, v_4 and v_5, as the values of the degree are all 1.0. We can also see the situation of the variable v_6. This variable is related to C_1 and C_2 to almost the same degree. This property can be seen in the data (given in figure 3.2). From this result, we can now investigate the validity of model (3.1).

Table 3.1. Fuzzy Clustering Result

Objects	C_1	C_2
o_1	0.75	0.25
o_2	0.64	0.36
o_3	0.88	0.12
o_4	0.83	0.17
o_5	0.12	0.88
o_6	0.06	0.94
o_7	0.04	0.96
o_8	0.03	0.97

Table 3.2. Fuzzy Cluster Loadings for 2-Way Data

Variables	C_1	C_2
v_1	1.00	0.00
v_2	0.00	1.00
v_3	0.00	1.00
v_4	0.00	1.00
v_5	0.00	1.00
v_6	0.49	0.51
v_7	1.00	0.00

Next, we use artificially created data shown in table 3.3 and figure 3.3. Table 3.3 shows values of eight objects, $o_1, \ldots, o_8$ with respect to three variables v_1, v_2, and v_3.

Table 3.3. Artificial Data

Objects	v_1	v_2	v_3
o_1	10	1	2
o_2	9	3	4
o_3	13	3	2
o_4	14	4	5
o_5	4	11	12
o_6	6	9	8
o_7	2	10	11
o_8	3	8	9

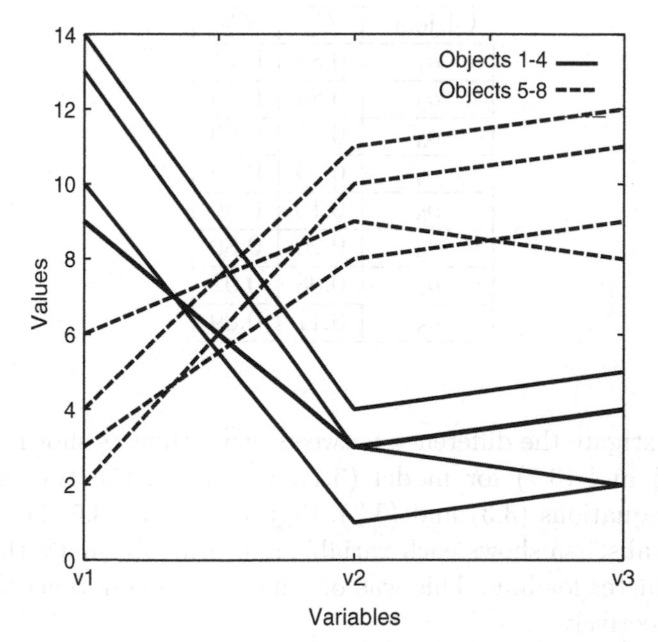

Fig. 3.3. Artificial Data

In figure 3.3, the abscissa shows each variable and the ordinate shows the values for each variable. The solid lines show the objects o_1, o_2, o_3 and o_4, and the dotted lines show the objects o_5, o_6, o_7 and o_8. From these, we can see the similarity of objects o_1, o_2, o_3 and o_4, and the similarity of objects o_5, o_6, o_7 and o_8. The feature of this data is that the first group, which consists of objects o_1, o_2, o_3, and o_4, has the property that the values of variable v_1 are large and the values of variables v_2 and v_3 are small. While the other group, which consists of objects o_5, o_6, o_7, and o_8, has the opposite property.

The result of fuzzy clustering using the FANNY algorithm [27] is shown in table 3.4. In this table, each value shows degree of belongingness of objects to each of the clusters, C_1 and C_2. We can see that objects o_1 to o_4 for the most part belong to cluster C_1 and objects o_5 to o_8 belong to cluster C_2. According to the feature of this data, which was created with the two clusters, this result is adaptable.

Table 3.4. Fuzzy Clustering Result

Objects	C_1	C_2
o_1	0.89	0.11
o_2	0.85	0.15
o_3	0.91	0.09
o_4	0.84	0.16
o_5	0.10	0.90
o_6	0.20	0.80
o_7	0.08	0.92
o_8	0.11	0.89

To investigate the difference between two estimates shown in equations (3.3) and (3.7) for model (3.1), we obtain the two estimates shown in equations (3.3) and (3.7). Figures 3.4 and 3.5 show the results. The abscissa shows each variable and ordinate shows the values of fuzzy cluster loading. This was obtained using equations (3.3) and (3.7), respectively.

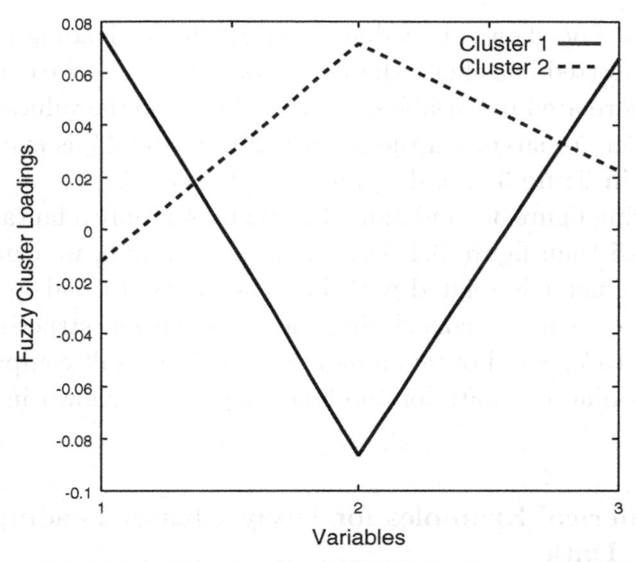

Fig. 3.4. Result of Fuzzy Cluster Loading in Equation (3.3)

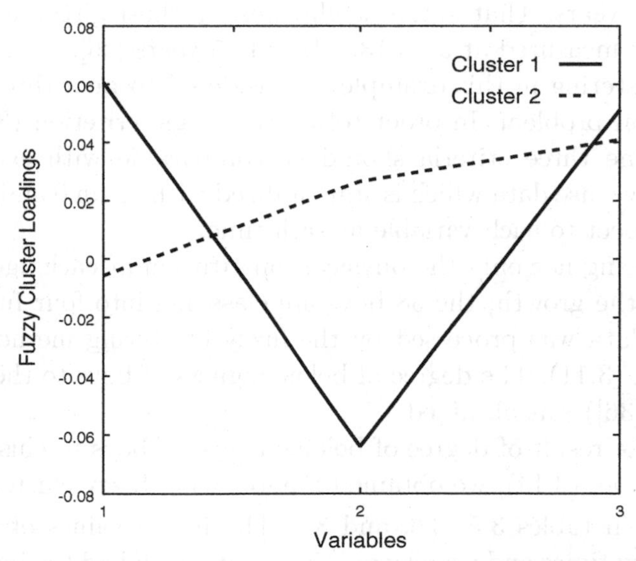

Fig. 3.5. Result of Fuzzy Cluster Loading in Equation (3.7)

The solid line shows the values of fuzzy cluster loading for cluster C_1 and the dotted line shows the values for cluster C_2. We can see that cluster C_1 is related to variables v_1 and v_3, because the values of degree of proportion are large. On the other hand, cluster C_2 is explained by variable v_2 in figure 3.4, and v_2 and v_3 in figure 3.5.

Comparing figure 3.4 and figure 3.5, there is a more adaptable result for figure 3.5 than figure 3.4 because from figure 3.3, we can see that the second cluster is related with both variables, v_2 and v_3. That is, the proposed estimate considering the classification structure shown in equation (3.7) could obtain a more reasonable result compared with the conventional estimate for the linear regression shown in equation (3.3).

3.3.2 Numerical Examples for Fuzzy Cluster Loading for 3-Way Data

We next use a numerical example which used 3-way data. This involved 38 boys and four observation on each boy were taken for three consecutive years. That is the height, weight, chest girth, and sitting height were measured at ages 13, 14, and 15 years [36].

The clustering in this example is considered to be a three criteria optimization problem. In order to use the single criterion (3.11), the values of the three criteria should be comparable with each other. Therefore, we use data which is standardized with mean 0 and variance 1, with respect to each variable at each time.

Considering not only the physical constitution at each age but the pattern of the growth, the 38 boys are classified into four fuzzy clusters. The data was processed by the fuzzy clustering method shown in equation (3.11). The degree of belongingness of boys to the clusters (shown in [36]) was obtained.

Using the result of degree of belongingness of boys to clusters, u_{ik}, and the model (3.14), we obtained the following fuzzy cluster loading $\eta_{ak}^{(t)}$ shown in tables 3.5, 3.6, and 3.7. The initial values of $\eta_{ak}^{(t)}$ were given twenty times and we selected the result which had the best fitness in equation (3.18).

Table 3.5. Fuzzy Cluster Loadings for Variables at the First Time Measurement (13 years old)

Variables	C_1	C_2	C_3	C_4
Height	0.02	0.31	0.21	0.46
Weight	0.67	0.02	0.01	0.30
Chest Girth	0.03	0.47	0.43	0.07
Sitting Height	0.25	0.17	0.37	0.22

Table 3.6. Fuzzy Cluster Loadings for Variables at the Second Time Measurement (14 years old)

Variables	C_1	C_2	C_3	C_4
Height	0.03	0.19	0.31	0.48
Weight	0.65	0.19	0.00	0.16
Chest Girth	0.02	0.35	0.48	0.15
Sitting Height	0.27	0.23	0.23	0.27

Table 3.7. Fuzzy Cluster Loadings for Variables at the Third Time Measurement (15 years old)

Variables	C_1	C_2	C_3	C_4
Height	0.09	0.24	0.26	0.41
Weight	0.65	0.21	0.02	0.11
Chest Girth	0.01	0.35	0.42	0.23
Sitting Height	0.22	0.16	0.31	0.31

From tables 3.5, 3.6, and 3.7, we can see in table 3.5, clusters C_2 and C_4 are high loadings with the variable height. However, in table 3.6, height is related with clusters C_3 and C_4. In table 3.7, clusters C_2, C_3, and C_4 are related to height. According to the result of centroids [36], we could see that cluster C_2 shows a poor physical constitution through all the years. C_3 and C_4 may be thought to represent the standard physical constitution. There is however a difference in the pattern of growth, namely cluster C_3 has the tendency of growth in both height and weight through all the years. Cluster C_4 contains the boys who grow through all the years but not so conspicuously. From this, the result of tables 3.5, 3.6, and 3.7 seem to show that at 13 years old, for these boys the height is just following the physical development. At 14 years old, the height is related with clusters C_3 and C_4 which may be thought of as growing factors. At the third time, the 15 year old boy's height is related with both physical development and growing factors.

Table 3.8 shows the result of equation (3.15). The initial values of η_{ak} were given twenty times and the result which had the best fitness shown in equation (3.19) was selected. We used the same result of degree of belongingness of boys to the clusters as we used previously in the analysis in tables 3.5, 3.6, and 3.7. The weights are assumed to be $w^{(t)} = \frac{1}{3}$, $t = 1, 2, 3$.

Table 3.8. Fuzzy Cluster Loadings for Variables

Variables	C_1	C_2	C_3	C_4
Height	0.01	0.25	0.27	0.47
Weight	0.69	0.09	0.00	0.22
Chest Girth	0.00	0.43	0.46	0.11
Sitting Height	0.27	0.18	0.29	0.26

From this result, we see that this result is very similar the results of tables 3.5, 3.6, and 3.7, so we could obtain similar interpretations of clusters C_1, C_2, C_3, and C_4. We can also see that the values of table 3.8 are similar to average values of tables 3.5, 3.6, and 3.7 for each of the values. From this, the result of equation (3.15) shows the general feature of the fuzzy clustering results over the times.

Next, we show an example using Landsat data which was observed over the Kushiro marsh-land. The value of the data shows the amount of reflected light from the ground with respect to six kinds of light for 75 pixels and these values were observed three times, August 1991, May 1992, and June 1992. This data is a 3-way data which consists of objects (pixels) $\times$ variables (light wavelengths) $\times$ times. The 3-way data, which is observed as the values of 75 objects with respect to six variables for three times, is denoted by equation (3.9). Here $n = 75$,

$p = 6$, $t = 3$ in equation (3.9). We next create the following 75×18 matrix:

$$X = (X^{(1)} X^{(2)} X^{(3)}). \tag{3.45}$$

In equation (3.45), the matrix X means that variables which are used at a particular time are independent from the variables at a different time. That is true even if the same variables are used at different times. For the first time, we get data from the mountain area, river area, and city area, which shows the first 25 pixels, the pixels in the range 26 - 50th pixels, and in the 51 - 75th pixel range, respectively. And for the second and the third times, we get the data for the pixels which correspond with pixels at the first time. Note that if the landscape was changed, then these pixels no longer indicate the three areas.

Using the data X in equation (3.45), we apply fuzzy c-means to obtain the result. Figure 3.6 shows the result of fuzzy c-means with $m = 1.25$. In figure 3.6, each axis shows each cluster and each symbol shows the pixel which is observed in the area of mountain ($\times$), river ($+$), and city ($*$) at the first time. The value of the coordinate for the symbol shows the degree of belongingness for each cluster. From this result, we can see that the observed data is clearly divided into three clusters which may show mountain area, river area, and the city. However, in this result, we cannot see movement or changing process on these three times. Therefore, we use model (3.1) under the conditions (3.16) and (3.17) in order to capture the latent effectiveness on the variables through the times for which the clusters are obtained.

Using the fuzzy clustering result shown in figure 3.6 and the obtained data (3.45), we apply the data to model (3.1). The result is shown in table 3.9. In the table 3.9, C_1, C_2, and C_3 are the three clusters in figure 3.6, and $v_a(t)$ shows a-th variable at time t. From table 3.9, we can find that at the first time, cluster C_1 (mountain) is characterized by variable v_1. However, at the second and the third times, the degree for the variable v_1 moved to cluster C_3 (city). This means that at the second and at the third times, the values of v_1 for the mountain pixels (at the first time) have become small.

A similar movement can be shown in cluster C_2 for a river. At the first time, cluster C_2 was characterized by variables v_4, v_5, and v_6. At the second time, variables v_5's and v_6's loading weights moved to the city area which is cluster C_3. We can see that the changing situation is more drastic from the first to the second time, as when comparing the change from the second to the third time. So, the structure of the landscape was drastically changed from mountain to city in this period. Also, the period from the first time to the second time is longer than the period from the second to the third time.

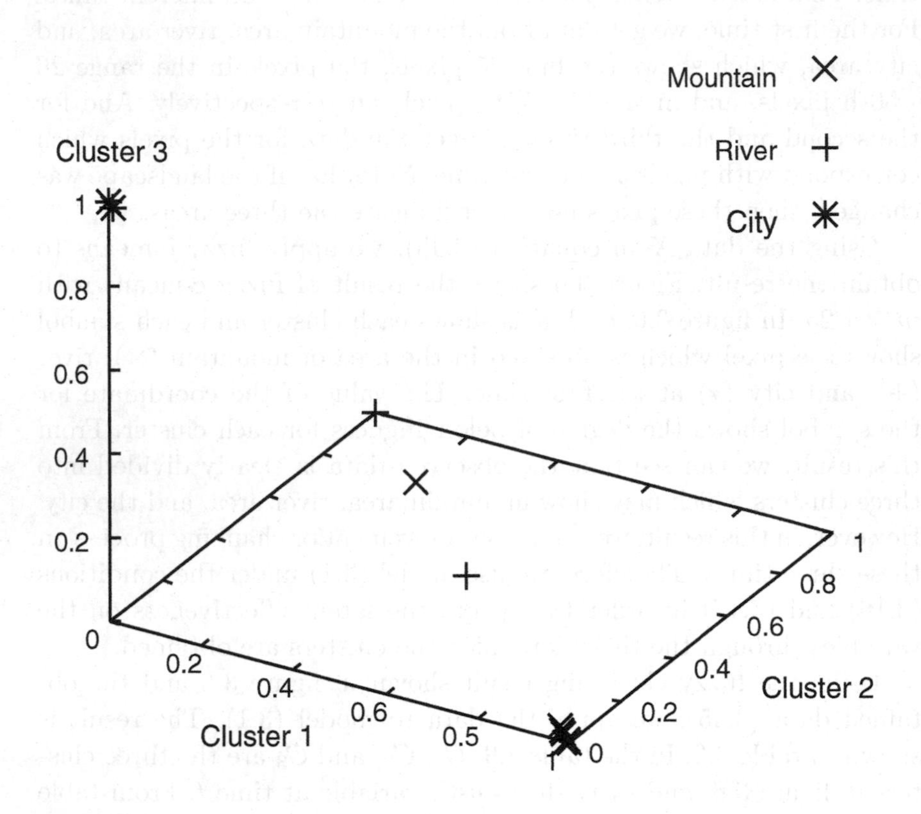

Fig. 3.6. Clustering of Landsat Data ($m = 1.25$)

Table 3.9. Cluster Loadings for Variables over Times

Variables	C_1	C_2	C_3
$v_1(1)$	1.00	0.00	0.00
$v_2(1)$	0.66	0.00	0.34
$v_3(1)$	0.00	0.00	1.00
$v_4(1)$	0.00	1.00	0.00
$v_5(1)$	0.00	1.00	0.00
$v_6(1)$	0.00	1.00	0.00
$v_1(2)$	0.00	0.00	1.00
$v_2(2)$	0.00	0.00	1.00
$v_3(2)$	0.00	0.00	1.00
$v_4(2)$	0.00	1.00	0.00
$v_5(2)$	0.00	0.00	1.00
$v_6(2)$	0.00	0.00	1.00
$v_1(3)$	0.00	0.00	1.00
$v_2(3)$	0.00	0.00	1.00
$v_3(3)$	0.00	0.00	1.00
$v_4(3)$	0.00	1.00	0.00
$v_5(3)$	0.00	1.00	0.00
$v_6(3)$	0.00	0.00	1.00

In order to test the validity of the result, we also examined this data in another way. The 3-way data was changed as follows:

$$X = \begin{pmatrix} X^{(1)} \\ X^{(2)} \\ X^{(3)} \end{pmatrix}. \tag{3.46}$$

Using the data (3.46), we apply fuzzy c-means with $m = 1.25$ shown in equation (1.3). In this case, we can obtain three fuzzy clustering results corresponding to the first, the second, and the third times. The merit of this idea (3.46) is that we can get the three results with respect to the same clusters through times. So, we can compare the results. The three results are shown in figures 3.7 - 3.9. Figures 3.7, 3.8, and 3.9 show the results at the first, the second, and the third times, respectively. In these figures 3.7 - 3.9, each axis shows a cluster and each symbol shows the pixel which is observed in the areas of mountain ($\times$), river ($+$), or city ($*$) at the first time. The value of the coordinate for the symbol shows the degree of belongingness for each cluster. Note that the axis for figures 3.7, 3.8, and 3.9 are identical to each other, because we obtained the results with respect to the same clusters. We are thus able to compare the results shown in these figures.

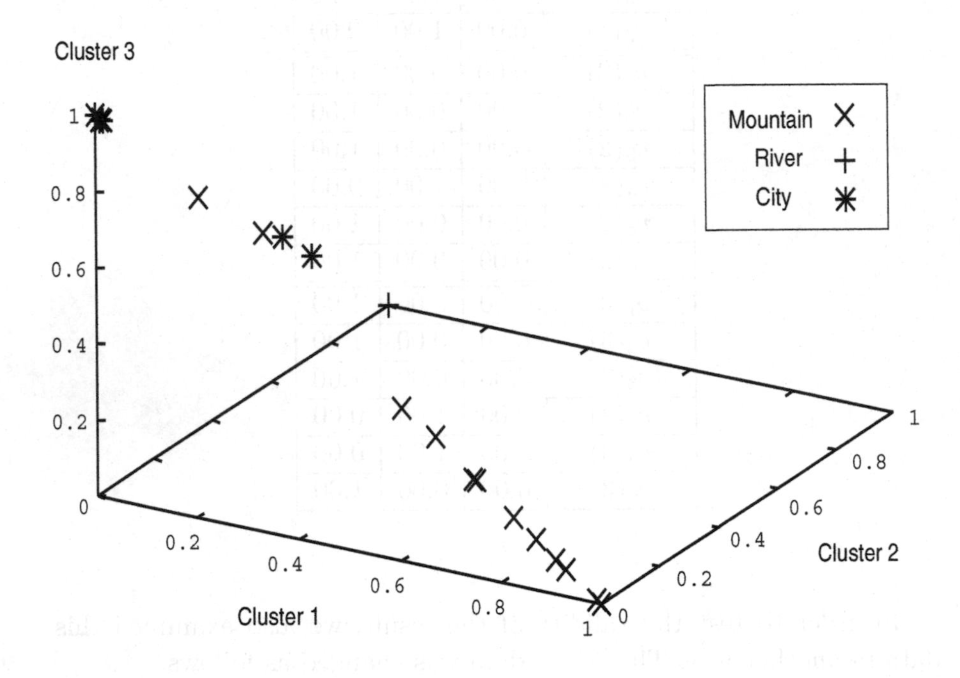

Fig. 3.7. Clustering of Landsat Data at the First Time ($m = 1.25$)

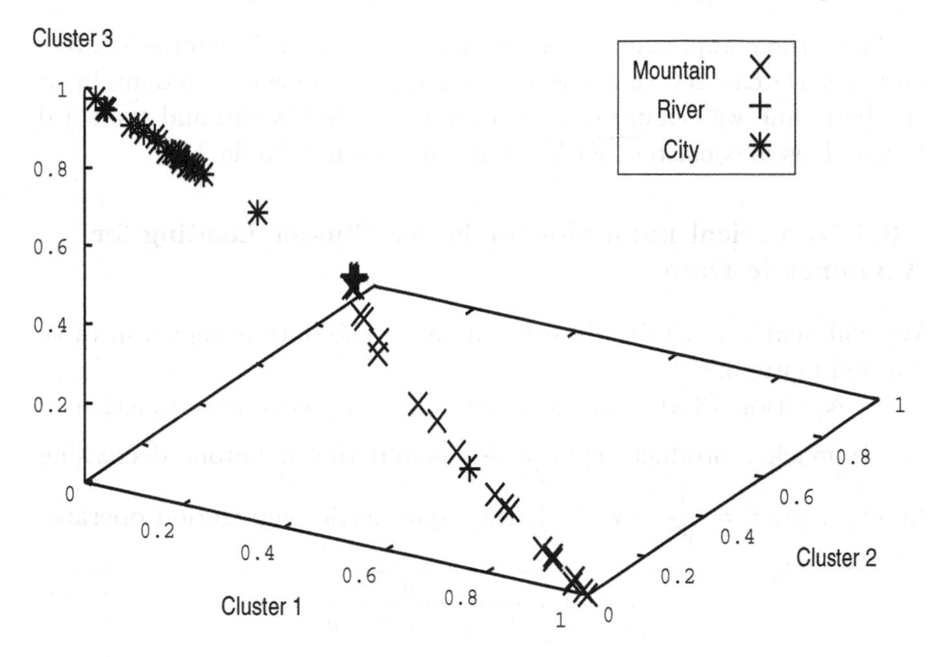

Fig. 3.8. Clustering of Landsat Data at the Second Time ($m = 1.25$)

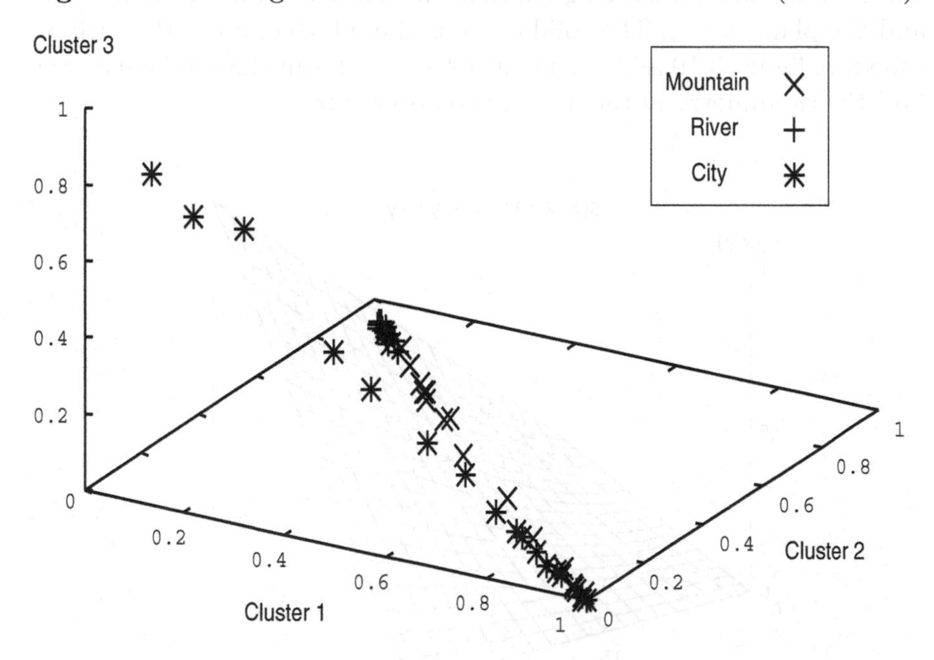

Fig. 3.9. Clustering of Landsat Data at the Third Time ($m = 1.25$)

From the comparison of the three results shown in figures 3.7, 3.8, and 3.9, we can see that the area which was shown as mountain at the first time was changed to a city area at the second and the third times. This is consistent with the result shown in table 3.9.

3.3.3 Numerical Examples for Fuzzy Cluster Loading for Asymmetric Data

We will next use artificially created data. The data is shown in table 3.3 and figure 3.3.

In equation (3.31), for instance, using the generator function of the Hamacher product, $f(x) = \dfrac{1-x}{x}$ and the monotone decreasing function $\phi(x) = \dfrac{1}{x^{\gamma}}$ $(\gamma > 0)$, the asymmetric aggregation operator obtained is

$$s(x, y) = \frac{x^{\gamma} y}{1 - y + x^{(\gamma-1)} y}. \tag{3.47}$$

This is shown in figure 3.10 where $\gamma = 2$. In figure 3.11, the dotted curve shows the intersecting curve of the surface shown in figure 3.10 and the plane $x = y$. The solid curve is the intersection of the surface shown in figure 3.10 and the plane $x + y = 1$. From the solid curve, we find the asymmetry of the aggregation operator.

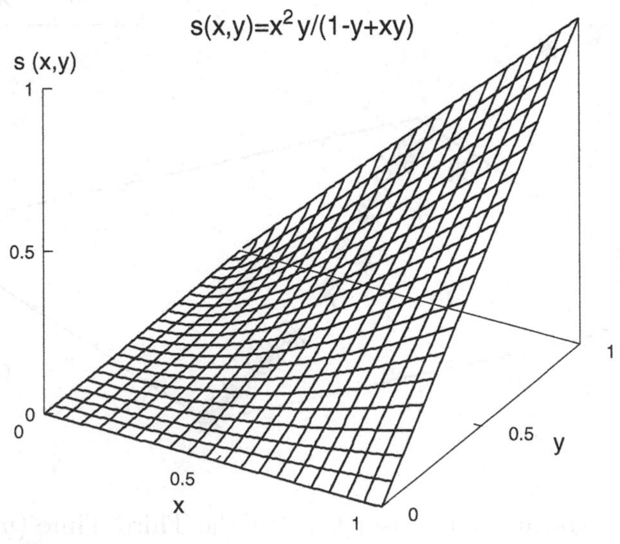

Fig. 3.10. Asymmetric Aggregation Operator

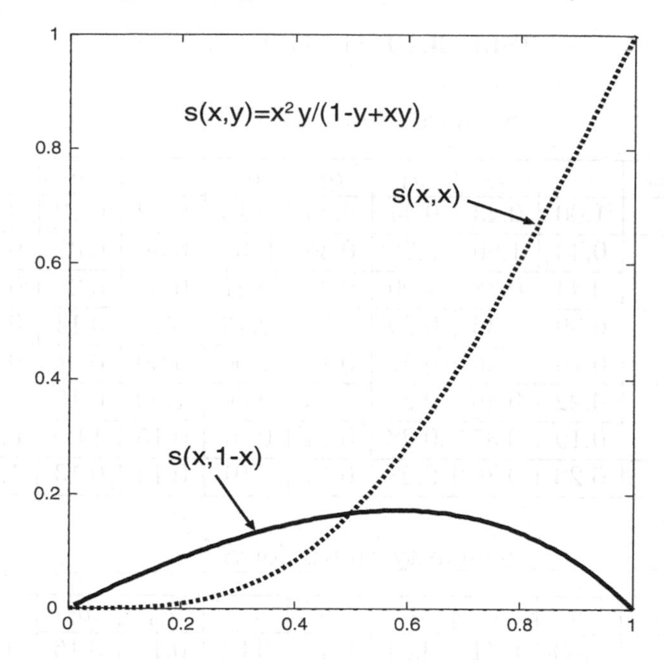

Fig. 3.11. Intersecting curves

Generally, $s(x, y)$ satisfies the inequality

$$s(x, y) \leq \rho(x, y)$$

by $f(x)+f(y) \leq f(x)+\phi(x)f(y)$, because $\phi(x) \geq 1$. Since the following inequality,

$$\sum_{k=1}^{K} s(u_{ik}, u_{jk}) \leq \sum_{k=1}^{K} \rho(u_{ik}, u_{jk}) \leq 1$$

is satisfied by equation (1.1), we assume the condition $0 \leq s_{ij} \leq 1$. Here the function ρ is assumed to satisfy the conditions of boundary, monotonicity, and symmetry described in section 3.1.3.

Using equation (3.47) ($\gamma = 2.0$) for the data shown in table 3.3, we obtain the 8×8 similarity matrixes for $S^{(U)}$ and $S^{(L)}$ of equation (3.32). They are shown in table 3.10. In this table, o_1 - o_8 show objects and each value shows the similarity between a pair of objects. $S^{(U)(L)}$ and $S^{(L)(U)}$ show the similarity matrixes which are created by equation (3.29) using the arithmetic mean. Therefore, in this case, $S^{(U)(L)}(= S^{(L)(U)})$ is a symmetric part of $S^{(U)}$ and $S^{(L)}$.

Table 3.10. Similarity Data

Similarity Matrix for $S^{(U)}$

Objects	o_1	o_2	o_3	o_4	o_5	o_6	o_7	o_8
o_1	1.00	0.43	0.44	0.39	0.15	0.22	0.19	0.26
o_2	0.43	1.00	0.38	0.36	0.36	0.36	0.37	0.36
o_3	0.44	0.38	1.00	0.29	0.27	0.28	0.28	0.28
o_4	0.39	0.36	0.29	1.00	0.13	0.13	0.14	0.14
o_5	0.15	0.36	0.27	0.13	1.00	0.00	0.01	0.01
o_6	0.22	0.36	0.28	0.13	0.00	1.00	0.15	0.14
o_7	0.19	0.37	0.28	0.14	0.01	0.15	1.00	0.13
o_8	0.26	0.36	0.28	0.14	0.01	0.14	0.13	1.00

Similarity Matrix for $S^{(L)}$

Objects	o_1	o_2	o_3	o_4	o_5	o_6	o_7	o_8
o_1	1.00	0.41	0.34	0.18	0.04	0.18	0.16	0.27
o_2	0.41	1.00	0.30	0.14	0.01	0.14	0.13	0.24
o_3	0.34	0.30	1.00	0.15	0.01	0.15	0.13	0.24
o_4	0.18	0.14	0.15	1.00	0.00	0.14	0.12	0.23
o_5	0.04	0.01	0.01	0.00	1.00	0.13	0.12	0.23
o_6	0.18	0.14	0.15	0.14	0.13	1.00	0.12	0.23
o_7	0.16	0.13	0.13	0.12	0.12	0.12	1.00	0.24
o_8	0.27	0.24	0.24	0.23	0.23	0.23	0.24	1.00

Similarity Matrix for $S^{(U)(L)} (= S^{(L)(U)} = (S^{(U)} + S^{(L)})/2)$

Objects	o_1	o_2	o_3	o_4	o_5	o_6	o_7	o_8
o_1	1.00	0.42	0.39	0.28	0.09	0.20	0.18	0.26
o_2	0.42	1.00	0.34	0.25	0.18	0.25	0.25	0.30
o_3	0.39	0.34	1.00	0.22	0.14	0.21	0.21	0.26
o_4	0.28	0.25	0.22	1.00	0.06	0.13	0.13	0.18
o_5	0.09	0.18	0.14	0.06	1.00	0.07	0.07	0.12
o_6	0.20	0.25	0.21	0.13	0.07	1.00	0.13	0.18
o_7	0.18	0.25	0.21	0.13	0.07	0.13	1.00	0.19
o_8	0.26	0.30	0.26	0.18	0.12	0.18	0.19	1.00

Table 3.11 shows the result of the fuzzy grade of $U^{(U)}$ in equation (3.33) and table 3.12 shows the result of $U^{(L)}$ in equation (3.33). In tables 3.11 and 3.12, C_1 and C_2 show the clusters and these clusters in table 3.11 are exactly the same as the clusters in table 3.12, so, we can compare the results. From the comparison of these results, we can see the difference of objects o_2, o_3, o_4, o_6, and o_7.

Table 3.11. Fuzzy Clustering Result for $U^{(U)}$

Objects	C_1	C_2
o_1	0.92	0.08
o_2	0.91	0.09
o_3	0.73	0.27
o_4	0.78	0.22
o_5	0.27	0.73
o_6	0.49	0.51
o_7	0.86	0.14
o_8	0.34	0.66

Table 3.12. Fuzzy Clustering Result for $U^{(L)}$

Objects	C_1	C_2
o_1	0.78	0.22
o_2	0.36	0.64
o_3	0.30	0.70
o_4	0.49	0.51
o_5	0.22	0.78
o_6	0.01	0.99
o_7	0.56	0.44
o_8	0.33	0.67

Tables 3.13 and 3.14 show the results of the fuzzy cluster loading for $U^{(U)}$ and $U^{(L)}$, respectively which are obtained by equation (3.34). In these tables, v_1, v_2, and v_3 show three variables and C_1 and C_2 are the two clusters. Note that the clusters C_1 and C_2 are the same as the clusters shown in tables 3.11 and 3.12. From the comparison of these matrixes, we can see the difference of the interpretation of clusters for the upper part and lower part of the asymmetric similarity data. In this case, these two matrixes show a similar tendency for the two clusters, but for v_2, we can see a larger difference when compared to the other two variables. From figure 3.3, we can see a clearer difference of the two groups for v_2, compared to v_1 and v_3. An explanation may be that the fuzzy cluster loading can capture the variables that have a large influence on the asymmetric structure of the asymmetric similarity data.

Table 3.13. Fuzzy Cluster Loading for $U^{(U)}$

Variables	C_1	C_2
v_1	0.53	0.47
v_2	0.41	0.59
v_3	0.58	0.42

Table 3.14. Fuzzy Cluster Loading for $U^{(L)}$

Variables	C_1	C_2
v_1	0.50	0.50
v_2	0.31	0.69
v_3	0.66	0.34

3.3.4 Numerical Examples for Fuzzy Clustering based the Weighted Regression Model

In order to search for the difference between the estimations of the weights in equation (3.38), we use the data shown in table 3.3 and figure 3.3. We compare the weight obtained by the kernel estimate shown in equation (3.43) using $r = 30$ with the weight obtained as an estimate by using fuzzy clustering results shown in equation (1.1).

That is we compare $\tilde{\boldsymbol{\beta}}_k$ in equation (3.39) and

$$\tilde{\hat{\boldsymbol{\beta}}}_k = (X' A_k^2 X)^{-1} X' A_k \boldsymbol{y}. \tag{3.48}$$

We treat the data with respect to v_1 in table 3.3 as the dependent variable for $\boldsymbol{y}$ and the data with respect to variables v_2 and v_3 as the independent variables for X in equation (3.38). Due to the comparison of $\tilde{\boldsymbol{\beta}}_k$ in equation (3.39) and $\tilde{\hat{\boldsymbol{\beta}}}_k$ in equation (3.48), we can calculate the following fitnesses:

$$R^U = \frac{1}{K} \sum_{k=1}^{K} (\boldsymbol{y} - \hat{U}_k^{-1} X \tilde{\boldsymbol{\beta}}_k)' (\boldsymbol{y} - \hat{U}_k^{-1} X \tilde{\boldsymbol{\beta}}_k). \tag{3.49}$$

$$R^A = \frac{1}{K} \sum_{k=1}^{K} (\boldsymbol{y} - A_k X \tilde{\hat{\boldsymbol{\beta}}}_k)' (\boldsymbol{y} - A_k X \tilde{\hat{\boldsymbol{\beta}}}_k). \tag{3.50}$$

Table 3.15 shows the values of regression coefficients obtained and table 3.16 shows the results of R^U and R^A shown in equations (3.49) and (3.50). From table 3.15, we can see that quite different results are obtained for equations (3.39) and (3.48). From the result shown in table 3.16, we can see that a better result is obtained by using the result of equation (3.39). That is, it is better to use fuzzy clustering for the weights rather than use the kernel estimate.

Table 3.15. Results of Regression Coefficients

Regression Coefficients	v_2	v_3
$\tilde{\boldsymbol{\beta}}_1$	3.53	0.55
$\tilde{\boldsymbol{\beta}}_2$	1.58	-1.03
$\tilde{\hat{\boldsymbol{\beta}}}_1$	-0.33	1.28
$\tilde{\hat{\boldsymbol{\beta}}}_2$	-0.27	0.89

Table 3.16. Comparison of the Weights

R^U	R^A
296.9	409.1

The result of equation (3.40) with $m = 2.0$ are compared with equation (3.42). From equations (3.41) and (3.44), the following fitnesses are used for the comparison:

$$S^U = \frac{1}{K}\sum_{k=1}^{K}(y - X\tilde{\tilde{\beta}}_k)'(\hat{U}_k^{-1})^2(y - X\tilde{\tilde{\beta}}_k). \tag{3.51}$$

$$S^A = \frac{1}{K}\sum_{k=1}^{K}(y - X\hat{\beta}_k)'A_k(y - X\hat{\beta}_k). \tag{3.52}$$

Table 3.17 shows the results of S^U and S^A in equations (3.51) and (3.52). We can see that the fitness obtained using fuzzy clustering is better.

Table 3.17. Comparison of the Weights

S^U	S^A
47.3	347.7

The data is the measurements of rainfall from 328 locations around Japan over a 12 month period and the average temperature for each location [69]. We take the data from 26 locations in the Hokkaido/Tohoku area and the Okinawa area. For the dependent variable, we use temperature data, and for the independent variables, we use data of rainfall over 12 months. The estimate of the degree of belongingness of a location to each cluster, $\hat{u}_{ik}$, is obtained by using the FANNY algorithm [27] for the rainfall data. The data was classified into two clusters. Figure 3.12 shows the result of the fuzzy clustering. In this figure, the abscissa shows location numbers and the ordinate shows the values of degree of belongingness of each location to each cluster. The solid line shows the fuzzy cluster grade for cluster C_1 and the dotted line shows the values for cluster C_2. The locations numbered from 1 to 14 show the locations in the Hokkaido/Tohoku area and the numbers 15 to 26 show the locations in the Okinawa area. From the result shown in figure 3.12, we can see clearly the two clusters. Cluster C_1 shows the Hokkaido/Tohoku area and the cluster C_2 shows the Okinawa area.

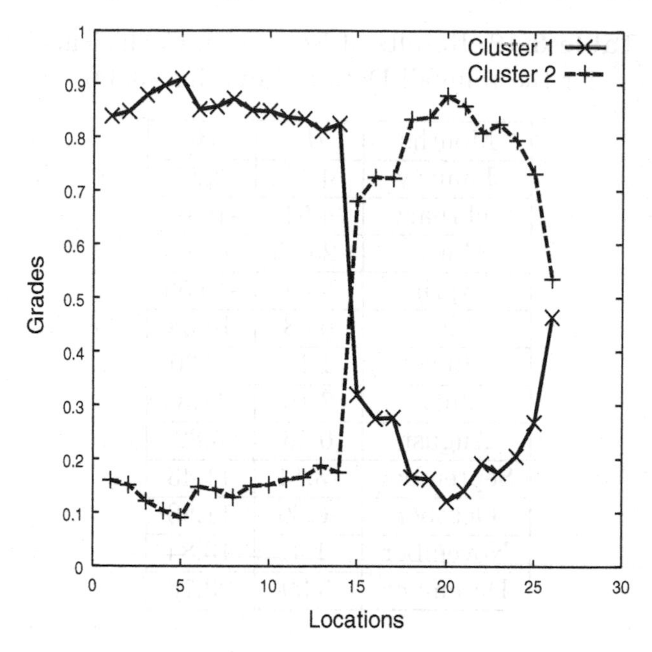

Fig. 3.12. Result of Fuzzy Clustering for the Rainfall Data

Using the result shown in figure 3.12 and equation (3.39), we obtain the result of regression coefficients shown in table 3.18 and figure 3.13. Using the kernel estimate shown in equation (3.43) with $r = 200$ and equation (3.48), we obtain the result of regression coefficients shown in table 3.19 and figure 3.14. In figure 3.13 and figure 3.14, the abscissa shows months and the ordinate shows the values of the regression coefficients of each cluster. The solid line shows the values of regression coefficients for cluster C_1 and the dotted line shows the values for cluster C_2.

Table 3.20 shows the fitness value for the results shown in figure 3.13 and figure 3.14. In this table, R^U and R^A are obtained by using equations (3.49) and (3.50). From figures 3.13 and 3.14, we can see that the tendency of these results might not have large differences, but table 3.20 shows that by the use of the fuzzy clustering result, we obtain a better fitness than with regression analysis.

Table 3.18. Results of Regression Coefficients for the Rainfall Data in Equation (3.39)

Months	$\tilde{\beta}_1$	$\tilde{\beta}_2$
January	4.19	32.05
February	4.54	-16.07
March	-23.45	-47.69
April	9.54	-19.55
May	6.58	16.08
June	1.16	-7.30
July	7.16	-11.35
August	0.56	5.82
September	-3.14	11.43
October	-0.30	11.90
November	-1.41	13.84
December	5.30	33.76

Table 3.19. Results of Regression Coefficients for the Rainfall Data in Equation (3.48)

Months	$\tilde{\hat{\beta}}_1$	$\tilde{\hat{\beta}}_2$
January	-32.04	102.55
February	-11.92	-97.46
March	-83.13	-97.15
April	21.81	-60.75
May	45.41	59.26
June	4.23	-43.17
July	50.60	-21.51
August	-27.80	-6.93
September	4.33	38.60
October	-28.37	40.67
November	6.31	28.68
December	45.06	129.16

Table 3.20. Comparison of the Fitness Values Obtained for the Rainfall Data

R^U	R^A
2380310	4317857

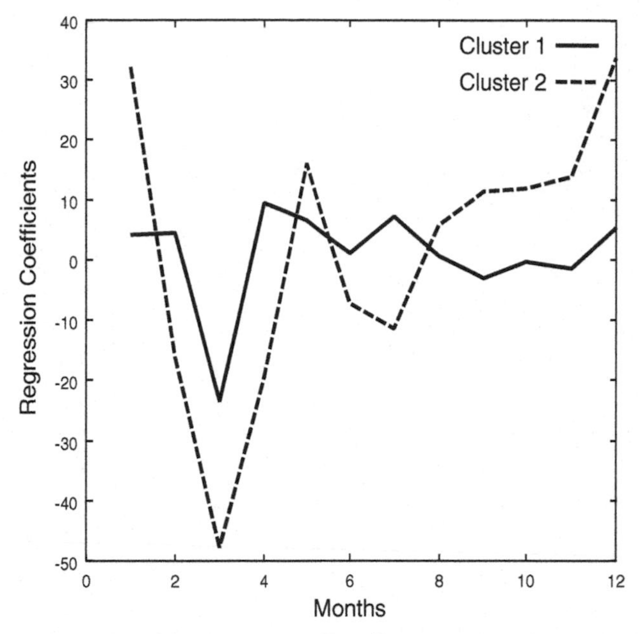

Fig. 3.13. Result of Regression Coefficients for the Rainfall Data in Equation (3.39)

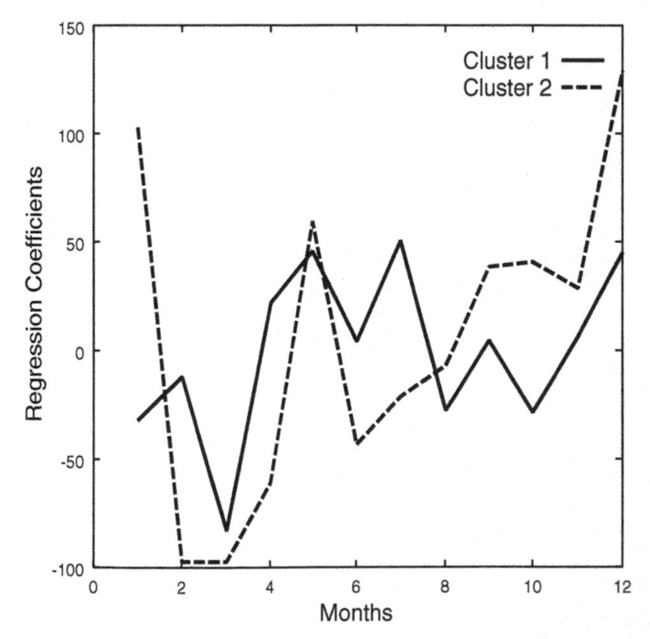

Fig. 3.14. Result of Regression Coefficients for the Rainfall Data in Equation (3.48)

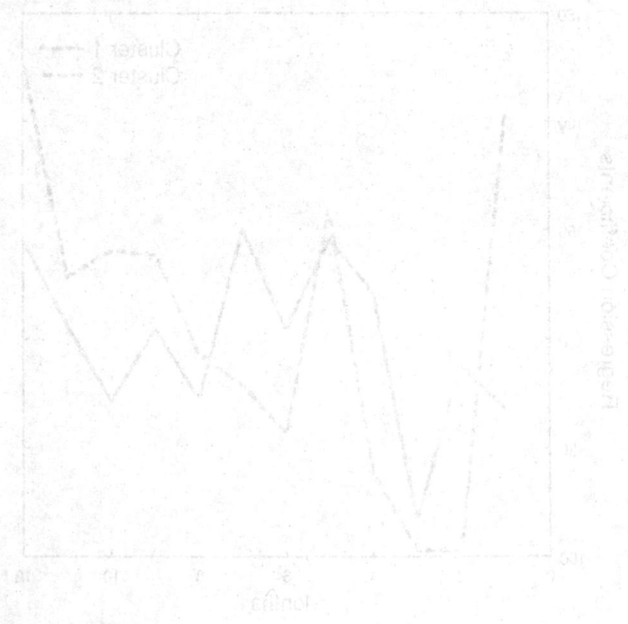

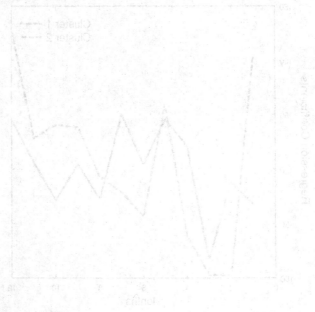

4. Kernel based Fuzzy Clustering

This chapter describes extended fuzzy cluster loadings obtained using the kernel method in a higher dimension space. It also shows the capability of kernel fuzzy cluster loading when used for the interpretation of the fuzzy clusters obtained from the fuzzy clustering result. The interpretation is estimated as a kernel fuzzy cluster loading which can show the relation between the obtained clusters and variables. We have mentioned fuzzy cluster loading in chapter 3. Conventional fuzzy cluster loading is estimated in the dimension of the observation space. However, kernel fuzzy cluster loading [47], [49] is estimated in the higher dimension space by applying the kernel method which is well known for a high discriminative performance in pattern recognition or classification, and this method can avoid the noise in the given observation. The higher dimension space is related to the data space nonlinearly, so it seems to be adaptable for the dynamic spatial data where the data performs irregularly over the variables and times. We illustrate the better performance of the kernel based fuzzy cluster loading with numerical examples.

4.1 Kernel Method

The kernel method is discussed in the context of support vector machines [8], [58]. Its advantages have been widely recognized in many areas. The essence of the kernel method is an arbitrary mapping from a lower dimension space to a higher dimension space. Note the mapping is an arbitrary mapping, so we do not need to find the mapping. This is referred to as the Kernel Trick.

Suppose there is an arbitrary mapping Φ:

$$\Phi\colon R^n \to F, \tag{4.1}$$

Mika Sato-Ilic and Lakhmi C. Jain: *Innovations in Fuzzy Clustering*, StudFuzz **205**, 89–104 (2006)
www.springerlink.com

where F is a higher dimension space than R^n. If a function ψ is the kernel function which is defined in R^n and ψ is a positive definite in R^n, then

$$\psi(\boldsymbol{x}, \boldsymbol{y}) = \Phi(\boldsymbol{x})' \Phi(\boldsymbol{y}), \quad \boldsymbol{x}, \ \boldsymbol{y} \in R^n.$$

Typical examples of the kernel function are as follows:

$$\psi(\boldsymbol{x}, \boldsymbol{y}) = \exp(-\frac{\|\boldsymbol{x} - \boldsymbol{y}\|}{2\sigma^2}). \tag{4.2}$$

$$\psi(\boldsymbol{x}, \boldsymbol{y}) = (\boldsymbol{x} \cdot \boldsymbol{y})^d. \tag{4.3}$$

$$\psi(\boldsymbol{x}, \boldsymbol{y}) = \tanh(\alpha(\boldsymbol{x} \cdot \boldsymbol{y}) + \beta). \tag{4.4}$$

Equation (4.2) is a gaussian kernel, equation (4.3) shows a polynomial kernel of degree d, and equation (4.4) is a sigmoid kernel. By the introduction of the kernel function, we can obtain the value of $\Phi(\boldsymbol{x})' \Phi(\boldsymbol{y})$ in F without finding the mapping Φ explicitly.

4.2 Kernel based Fuzzy Cluster Loading Models

From equation (3.7), we can obtain the following:

$$\begin{aligned} \tilde{\tilde{\boldsymbol{\eta}}}_k &= (X' \hat{U}_k^2 X)^{-1} X' \hat{U}_k \mathbf{1} \\ &= ((\hat{U}_k X)' (\hat{U}_k X))^{-1} (\hat{U}_k X)' \mathbf{1} \\ &\equiv (C_{(k)}' C_{(k)})^{-1} C_{(k)}' \mathbf{1}, \end{aligned} \tag{4.5}$$

where $C_{(k)} = (c_{ia(k)}), \ c_{ia(k)} \equiv \hat{u}_{ik}^{-1} x_{ia}, \ i = 1, \cdots, n, \ a = 1, \cdots, p.$

Using $\boldsymbol{c}_{a(k)} = \begin{pmatrix} c_{1a(k)} \\ \vdots \\ c_{na(k)} \end{pmatrix}$, we can represent equation (4.5) as follows:

$$\tilde{\tilde{\boldsymbol{\eta}}}_k = (\boldsymbol{c}_{a(k)}' \boldsymbol{c}_{b(k)})^{-1} (\boldsymbol{c}_{a(k)}' \mathbf{1}), \ a, b = 1, \cdots, p, \tag{4.6}$$

where $C_{(k)}' C_{(k)} = (\boldsymbol{c}_{a(k)}' \boldsymbol{c}_{b(k)}), \ C_{(k)}' \mathbf{1} = (\boldsymbol{c}_{a(k)}' \mathbf{1}), \ a, b = 1, \cdots, p$. Then we consider the following mapping Φ:

$$\Phi\colon R^n \to F, \quad \boldsymbol{c}_{a(k)} \in R^n. \tag{4.7}$$

From equations (4.6) and (4.7), the fuzzy cluster loading in F is as follows:

$$\hat{\boldsymbol{\eta}}_k = (\Phi(\boldsymbol{c}_{a(k)})'\Phi(\boldsymbol{c}_{b(k)}))^{-1}(\Phi(\boldsymbol{c}_{a(k)})'\Phi(\mathbf{1})), \quad a, b = 1, \cdots, p. \qquad (4.8)$$

In equation (4.8), $\hat{\boldsymbol{\eta}}_k$ shows the fuzzy cluster loading in F. Using the kernel representation $\psi(\boldsymbol{x}, \boldsymbol{y}) = \Phi(\boldsymbol{x})'\Phi(\boldsymbol{y})$ mentioned in the above section, equation (4.8) is rewritten as follows:

$$\hat{\boldsymbol{\eta}}_k = (\psi(\boldsymbol{c}_{a(k)}, \boldsymbol{c}_{b(k)}))^{-1}(\psi(\boldsymbol{c}_{a(k)}, \mathbf{1})), \ a, b = 1, \cdots, p. \qquad (4.9)$$

From this, using the kernel method, we estimate the fuzzy cluster loading in F and call this the kernel based fuzzy cluster loading for 2-way data.

For equation (3.14), we formulate the kernel fuzzy regression for 3-way data to obtain the fuzzy cluster loading for 3-way data in a higher dimension space as follows:

$$\tilde{\boldsymbol{\eta}}_k^{(t)} = (\psi(\boldsymbol{c}_{a(k)}^{(t)}, \boldsymbol{c}_{b(k)}^{(t)}))^{-1}(\psi(\boldsymbol{c}_{a(k)}^{(t)}, \mathbf{1})), \quad a, b = 1, \cdots, p,$$

where

$$(\boldsymbol{c}_{a(k)}^{(t)}) = \begin{pmatrix} c_{1a(k)}^{(t)} \\ \vdots \\ c_{na(k)}^{(t)} \end{pmatrix}, \quad c_{ia(k)}^{(t)} \equiv \hat{u}_{ik}^{-1} x_{ia}^{(t)},$$

$$i = 1, \cdots, n, \ a = 1, \cdots, p, \ k = 1, \cdots, K, \ t = 1, \cdots T.$$

From equation (3.37), $\tilde{\boldsymbol{\eta}}_{dk}^{(c)}$ can be shown to be as follows:

$$\tilde{\boldsymbol{\eta}}_{dk}^{(c)} = ((G_{dk}^{(c)})' G_{dk}^{(c)})^{-1}(G_{dk}^{(c)})'\mathbf{1}, \qquad (4.10)$$

where $G_{dk}^{(c)} = (g_{i^{(c)}d_{a(k)}})$, $g_{i^{(c)}d_{a(k)}} \equiv \hat{u}_{i^{(c)}d_k}^{-1} x_{i^{(c)}a}$, $i = 1, \cdots, n$, $a = 1, \cdots, p$. We rewrite equation (4.10) as:

$$\tilde{\boldsymbol{\eta}}_{dk}^{(c)} = ((\boldsymbol{g}_{a(k)}^{(c)d})' \boldsymbol{g}_{b(k)}^{(c)d})^{-1}((\boldsymbol{g}_{a(k)}^{(c)d})'\mathbf{1}), \ a, b = 1, \cdots, p, \qquad (4.11)$$

where

$$(\boldsymbol{g}_{a(k)}^{(c)d}) = \begin{pmatrix} g_{1^{(c)}d_{a(k)}} \\ \vdots \\ g_{n^{(c)}d_{a(k)}} \end{pmatrix},$$

$$(G_{dk}^{(c)})' G_{dk}^{(c)} = ((\boldsymbol{g}_{a(k)}^{(c)d})' \boldsymbol{g}_{b(k)}^{(c)d}), \ (G_{dk}^{(c)})'\mathbf{1} = ((\boldsymbol{g}_{a(k)}^{(c)d})'\mathbf{1}), \ a, b = 1, \cdots, p.$$

Then we introduce the mapping Φ in equation (4.1) for $\boldsymbol{g}_{a(k)}^{(c)d} \in R^n$. The kernel based fuzzy cluster loading in F for 3-way asymmetric data is defined as:

$$\hat{\boldsymbol{\eta}}_{dk}^{(c)} = (\Phi(\boldsymbol{g}_{a(k)}^{(c)d})' \Phi(\boldsymbol{g}_{b(k)}^{(c)d}))^{-1} (\Phi(\boldsymbol{g}_{a(k)}^{(c)d})' \Phi(\mathbf{1})), \quad a, b = 1, \cdots, p.$$

Using the kernel representation, $\hat{\boldsymbol{\eta}}_{dk}^{(c)}$ is obtained as:

$$\hat{\boldsymbol{\eta}}_{dk}^{(c)} = (\psi(\boldsymbol{g}_{a(k)}^{(c)d}, \boldsymbol{g}_{b(k)}^{(c)d}))^{-1} (\psi(\boldsymbol{g}_{a(k)}^{(c)d}, \mathbf{1})), \quad a, b = 1, \cdots, p. \tag{4.12}$$

4.3 Inner Product Space of Kernel Fuzzy Cluster Loading

From the definition of the kernel fuzzy cluster loading shown in equation (4.9), we can see that kernel fuzzy cluster loading depends on the result of fuzzy clustering, that is $\hat{u}_{ik}$. Also, the essential difference of the kernel fuzzy cluster loading and the fuzzy cluster loading is the difference between the product $\psi(\boldsymbol{c}_{a(k)}, \boldsymbol{c}_{b(k)}) \equiv \Phi(\boldsymbol{c}_{a(k)})' \Phi(\boldsymbol{c}_{b(k)})$ in F and the product $\boldsymbol{c}_{a(k)}' \boldsymbol{c}_{b(k)}$ in R^n. Since $\boldsymbol{c}_{a(k)} = \begin{pmatrix} c_{1a(k)} \\ \vdots \\ c_{na(k)} \end{pmatrix}$, $c_{ia(k)} \equiv \hat{u}_{ik}^{-1} x_{ia}$, these products are closely related to the result of fuzzy clustering. The result of fuzzy clustering is the degree of belongingness of the objects to the clusters, in other words $\hat{u}_{ik}$. Therefore, we investigate the property of these products for the change of the fuzzy clustering result. From $C_{(k)}' C_{(k)} = (\boldsymbol{c}_{a(k)}' \boldsymbol{c}_{b(k)}), \ C_{(k)}' \mathbf{1} = (\boldsymbol{c}_{a(k)}' \mathbf{1}), \ a, b = 1, \cdots, p$,

$$\boldsymbol{c}_{a(k)}' \boldsymbol{c}_{b(k)} = (\hat{u}_{1k}^{-1})^2 x_{1a} x_{1b} + \cdots + (\hat{u}_{nk}^{-1})^2 x_{na} x_{nb} = \sum_{i=1}^{n} (\hat{u}_{ik}^{-1})^2 x_{ia} x_{ib}. \tag{4.13}$$

If ψ is the polynomial kernel with degree 2 (see equation (4.3) with $d = 2$), then

$$\psi(\boldsymbol{c}_{a(k)}, \boldsymbol{c}_{b(k)}) = \Phi(\boldsymbol{c}_{a(k)})' \Phi(\boldsymbol{c}_{b(k)}) = (\boldsymbol{c}_{a(k)}' \boldsymbol{c}_{b(k)})^2 = (\sum_{i=1}^{n} (\hat{u}_{ik}^{-1})^2 x_{ia} x_{ib})^2. \tag{4.14}$$

For fixed a, b, when $n = 2$, we can simplify, equations (4.13) and (4.14) respectively as follows:

$$\boldsymbol{c}_{a(k)}^{'}\boldsymbol{c}_{b(k)} = w_1 x + w_2 y, \tag{4.15}$$

$$\psi(\boldsymbol{c}_{a(k)}, \boldsymbol{c}_{b(k)}) = \Phi(\boldsymbol{c}_{a(k)})^{'}\Phi(\boldsymbol{c}_{b(k)}) = (w_1 x + w_2 y)^2, \tag{4.16}$$

where,

$$w_i \equiv (\hat{u}_{ik}^{-1})^2, \ x \equiv x_{1a}x_{1b}, \ y \equiv x_{2a}x_{2b}, \ i = 1, 2.$$

Table 4.1 shows the changing situation of both equations (4.15) and (4.16) according to the change of the fuzzy clustering result. That is to the values of w_1 and w_2.

Table 4.1. Comparison between Equations (4.15) and (4.16) according to the Change in the Fuzzy Clustering Result (w_1 and w_2)

w_1	w_2	Product in R^n shown in (4.15)	Product in F shown in (4.16)
1.0	0.0	x	x^2
0.0	1.0	y	y^2
0.5	0.5	$0.5(x+y)$	$0.25(x+y)^2$
0.3	0.7	$0.3x + 0.7y$	$(0.3x + 0.7y)^2$

Figure 4.1 shows the value of equation (4.15) in R^n with respect to x and y where $w_1 = w_2 = 0.5$. Figure 4.2 is the value of equation (4.16) in F with respect to x and y where $w_1 = w_2 = 0.5$. In this case, the clustering structure is homogeneous compared to other cases in table 4.1. This is because $w_1 = w_2 = 0.5$.

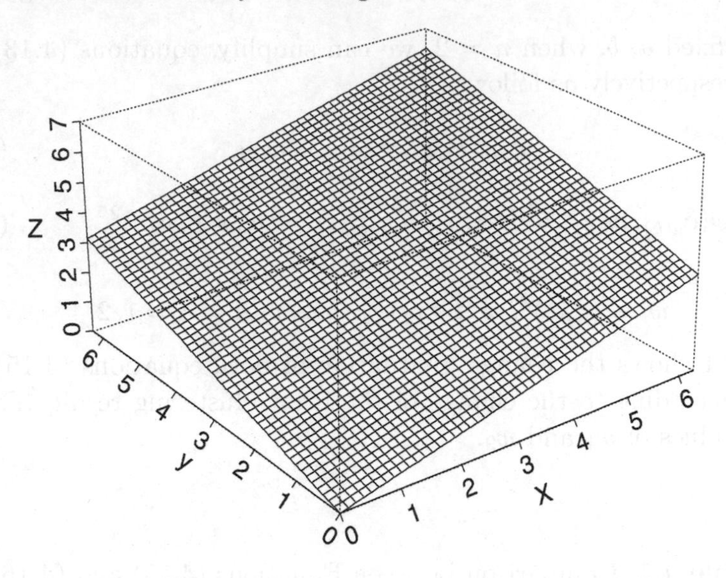

Fig. 4.1. Equation (4.15) in R^n when $w_1 = w_2 = 0.5$

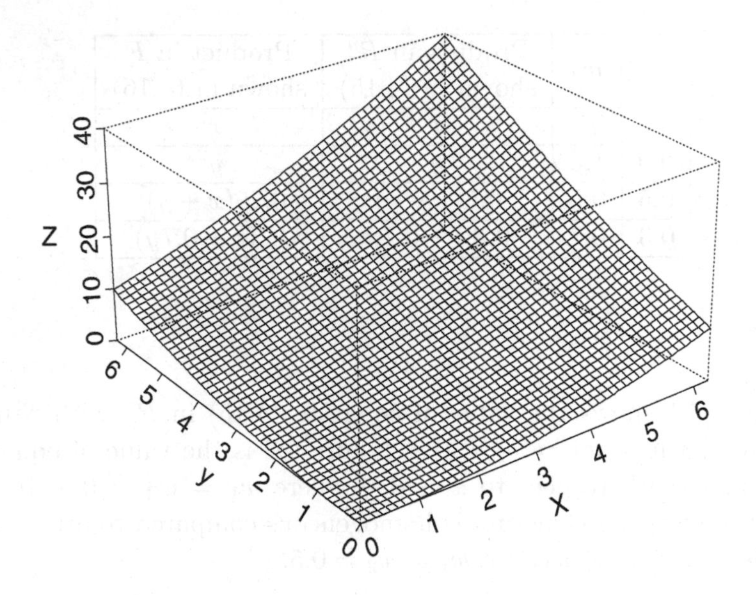

Fig. 4.2. Equation (4.16) in F when $w_1 = w_2 = 0.5$

Figure 4.3 shows the intersecting lines between the plane or the surface shown in figures 4.1 and 4.2 and the plane $x + y = 1$, respectively. In figure 4.3, the solid line shows the intersecting line between the plane in figure 4.1 and the plane $x + y = 1$ and the dotted line shows the intersecting line between the surface in figure 4.2 and the plane $x + y = 1$.

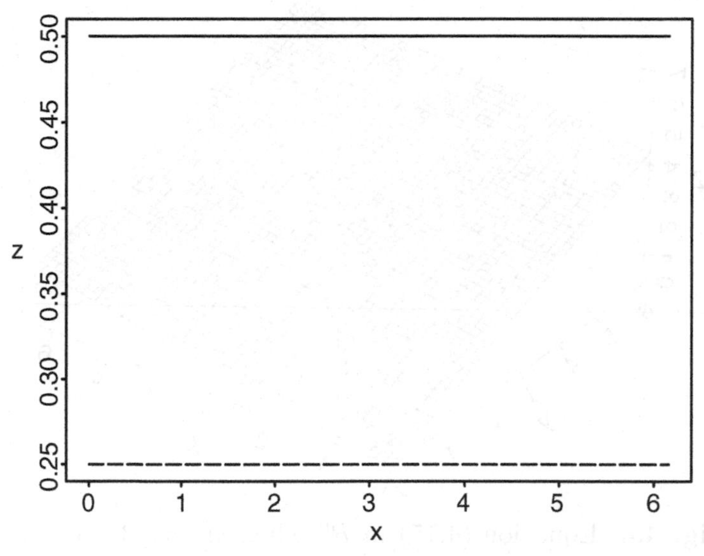

Fig. 4.3. Intersecting Lines

Figure 4.4 shows the value of equation (4.15) in R^n with respect to x and y in the case of $w_1 = 0.3$ and $w_2 = 0.7$. Figure 4.5 is the value of equation (4.16) in F with respect to x and y in the case of $w_1 = 0.3$ and $w_2 = 0.7$. In this case, the clustering structure is heterogeneous compared to the other cases in table 4.1, because $w_1 = 0.3$ and $w_2 = 0.7$.

Figure 4.6 shows the intersecting line and curve between the plane or the surface shown in figures 4.4 and 4.5 and the plane $x + y = 1$, respectively. In figure 4.6, the solid line shows the intersection between the plane in figure 4.4 and the plane $x + y = 1$. The dotted line shows the intersecting curve between the surface in figure 4.5 and the plane $x + y = 1$. Comparing figures 4.3 and 4.6, we can see more complex differences between the two products in figure 4.6. That is, if the clustering result is able to capture the heterogeneous structure of the given data, then the difference between the two products is more sensitively captured, that is when compared to the case where the cluster structure of the data is homogeneous. In other words, if the data has a structure where it can be classified into a small number of clusters, then the difference tends to be monotonic with respect to the value of the observation. However, if the data is classified into a large number of clusters, the difference is more complex.

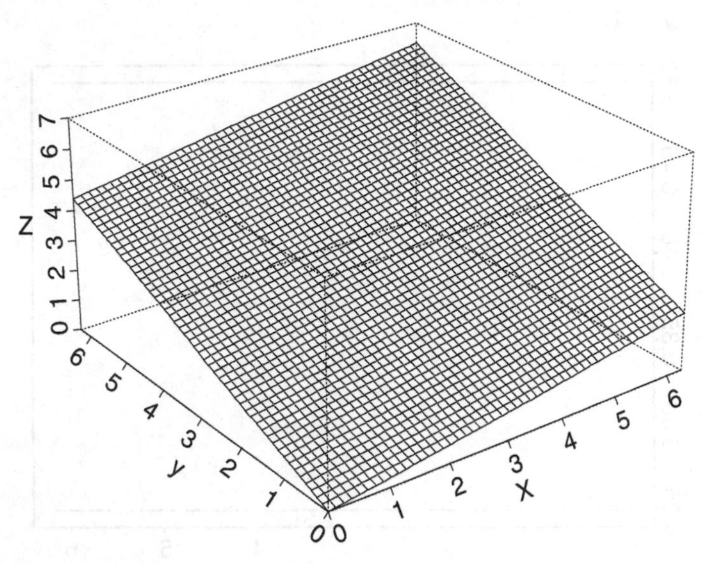

Fig. 4.4. Equation (4.15) in R^n when $w_1 = 0.3, w_2 = 0.7$

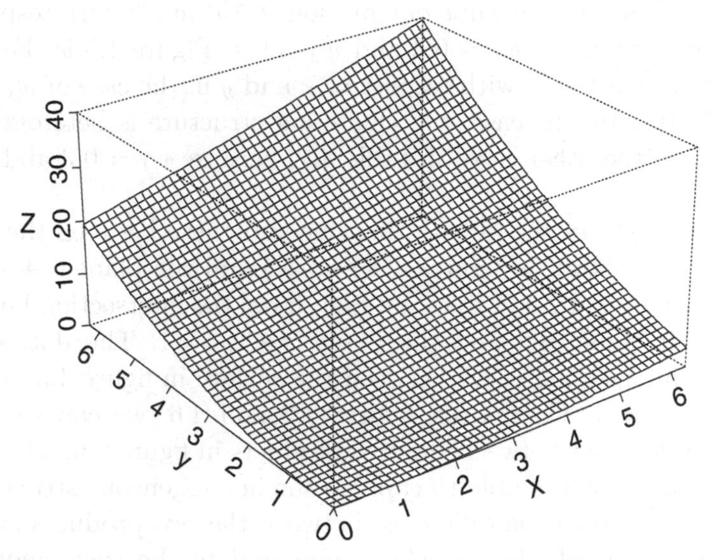

Fig. 4.5. Equation (4.16) in F when $w_1 = 0.3, w_2 = 0.7$

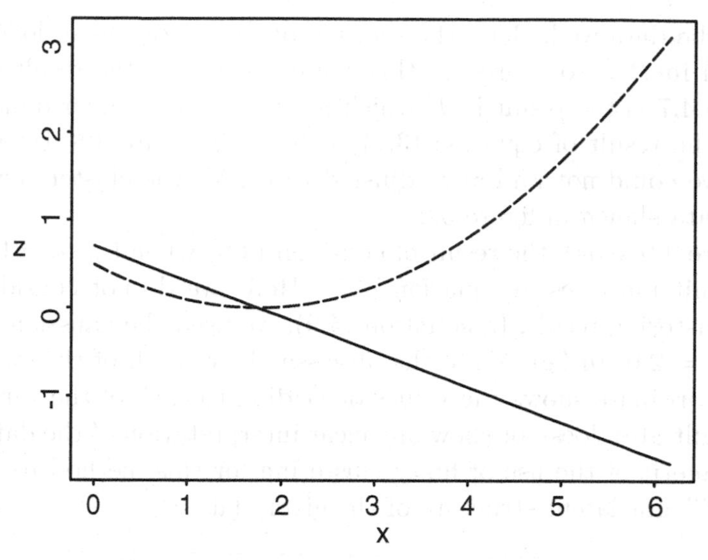

Fig. 4.6. Intersecting Line and Curve

4.4 Numerical Examples of Kernel based Fuzzy Cluster Loading Models

We first use artificially created data in order to get the validity of equation (4.9). The data is shown in figure 3.2. The result of fuzzy c-means using $m = 2.0$ (given in equation (1.3)) is shown in table 3.1. In order to find the properties of clusters C_1 and C_2, we obtain the estimate of the fuzzy cluster loading shown in equation (3.7) and the kernel fuzzy cluster loading shown in equation (4.9). Figure 4.7 shows the result of the kernel fuzzy cluster loading. In figure 4.7, the abscissa shows each variable and the ordinate shows the values of the fuzzy cluster loading obtained by using equation (4.9). For this we used the gaussian kernel shown in equation (4.2) and a value of $\sigma = 2.0$. The solid line shows the values of the fuzzy cluster loading for cluster C_1 and the dotted line shows the values for cluster C_2. From this result, we can see that cluster C_1 is related to variables v_1 and v_7, because the values of degree of proportion are large. On the other hand, cluster C_2 is explained by variables v_2 to v_6, as the values of the fuzzy cluster loading for these variables are large. This is a property which we can see in the data in figure 3.2.

The conventional estimation method using weighted regression analysis which is shown in equation (3.7), could not extract the structure of the data well. Here, the solution of the fuzzy cluster loading is obtained in R^n. Compared to this result, note that the result shown in figure 4.7 is the result in F which is the mapped higher dimension space. The result of equation (3.7) is shown in figure 4.8. Using this figure, we could not find any adjusted result for the cluster structure of the data shown in figure 3.2.

Figure 4.9 shows the result of equation (4.9) when $\hat{U}_k$, $k = 1, 2$ are $n \times n$ unit matrixes in equation (4.5). Here, we do not consider the fuzzy clustering result. In equation (4.9), we used the gaussian kernel where $\sigma = 2.0$. In figure 4.9, the abscissa shows each of the variables and the ordinate shows the values of loading to each of the variables. This result also does not show any clear interpretation of the data. We see the merit of the use of fuzzy clustering for this method to obtain the significant latent structure of the given data.

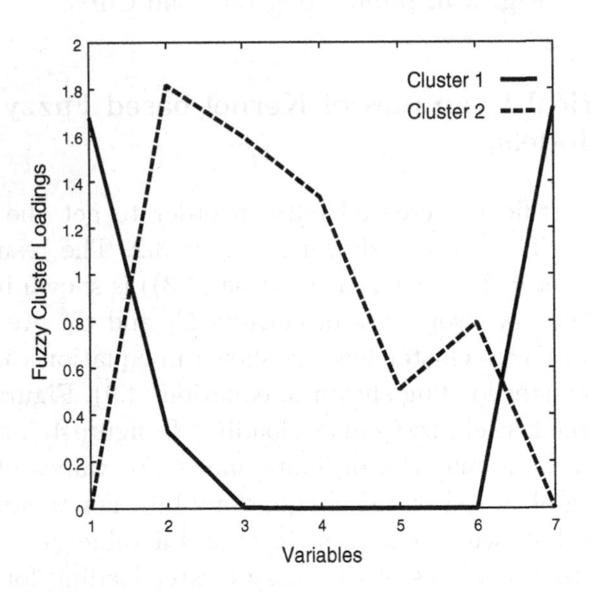

Fig. 4.7. Result of Fuzzy Cluster Loading in F using Kernel Fuzzy Regression based on Fuzzy Clustering using the Gaussian Kernel

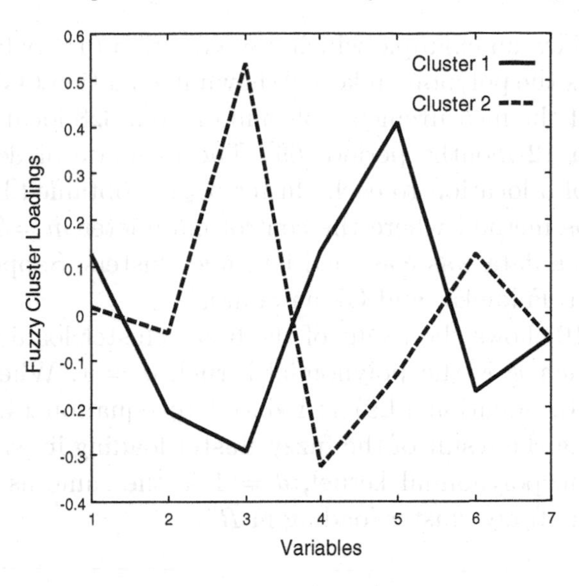

Fig. 4.8. Result of Fuzzy Cluster Loading in R^n using Fuzzy Regression based on Fuzzy Clustering

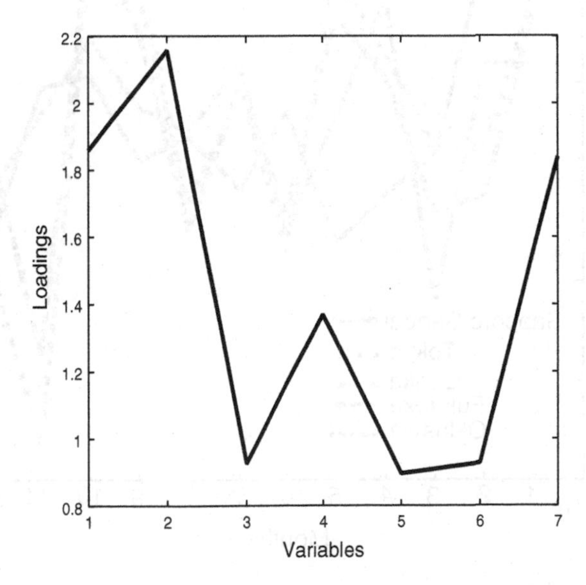

Fig. 4.9. Result of Loading in F using Kernel Fuzzy Regression using the Gaussian Kernel

We now show an example which uses equation (4.9) where the kernel function is the polynomial kernel shown in equation (4.3). The data is made up of the measurements of rainfall from 328 locations around Japan over a 12 months period [69]. The estimate of degree of belongingness of a location to each cluster, $\hat{u}_{ik}$, is obtained by using the fuzzy c-means method where the control parameter $m = 2.0$ in equation (1.3). The data was classified into five clusters, Sapporo/Sendai, Tokyo, Osaka, Fukuoka, and Okinawa areas.

Figure 4.10 shows the result of the fuzzy cluster loadings in equation (4.9) when ψ is the polynomial kernel, $d = 1$. When $d = 1$ in equation (4.3), equation (4.9) is reduced to equation (4.6). That is why, obtaining the result of the fuzzy cluster loading in equation (4.9) when ψ is the polynomial kernel, $d = 1$ is the same as finding the solution of the fuzzy cluster loading in R^n.

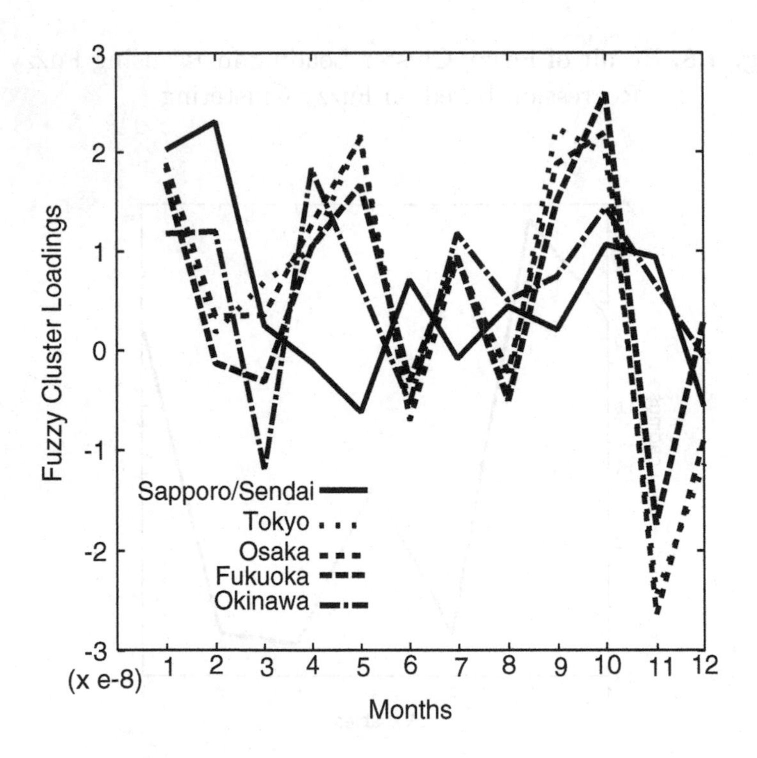

Fig. 4.10. Fuzzy Cluster Loading for the Rainfall Data using Polynomial Kernel ($d = 1$)

In figure 4.10, the abscissa shows each month as the variable and the ordinate is the value of the fuzzy cluster loadings. Each line shows each cluster. From this figure, we see that the Sapporo/Sendai area has an opposite situation to the other areas, especially in the month of February. The Sapporo/Sendai area does not have as much rainfall, instead they receive snow due to lower temperatures.

Figure 4.11 shows the result of equation (4.9) using the polynomial kernel when $d = 2$. Here, the estimated fuzzy cluster loading is a solution obtained in F (mapped higher dimension space). From this figure, we can see the same feature for February. That is, Sapporo/Sendai are remarkably different when compared to the other four areas.

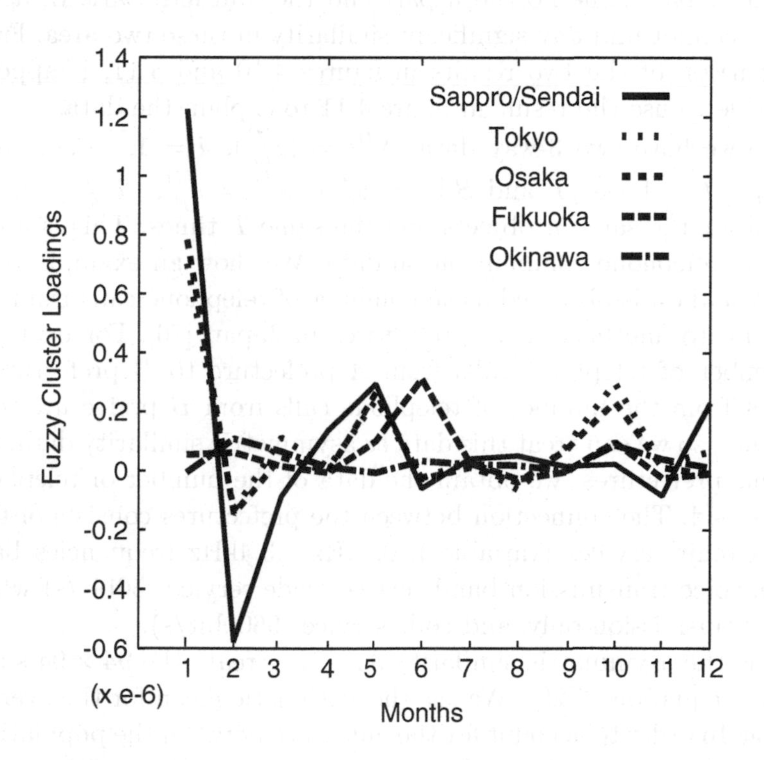

Fig. 4.11. Fuzzy Cluster Loading for the Rainfall Data using Polynomial Kernel ($d = 2$)

We can see clearer properties in figure 4.11, when comparing to the result shown in figure 4.10. For example, in May, Sapporo/Sendai has clearly different features to the other four areas in figure 4.10. In figure 4.11, we see that the difference is small and Sapporo/Sendai, Tokyo, and Osaka are similar to each other. The next smaller value is Fukuoka and the smallest value is Okinawa. Since these five areas are located in order from north to south according to the list, Sapporo/Sendai, Tokyo, Osaka, Fukuoka, Okinawa, the values for the fuzzy cluster loading in May are also arranged in order from north to south. The result seems to be reasonable. In November, we see similarity between Sapporo/Sendai and Okinawa in figure 4.10. This is difficult to explain because the location of these two areas are quite different. That is, the northern part and the southern part. In figure 4.11, we cannot find any significant similarity in these two area. From a comparison of the two results in figures 4.10 and 4.11, it appears reasonable to use the result in figure 4.11 to explain the data.

Here we have two 3-way data, $X^{(t)} = (x_{ia}^{(t)})$, $i = 1, \cdots, n$, $a = 1, \cdots, p$, $t = 1, \cdots, T$ and $S^{(t)} = (s_{ij}^{(t)})$, $s_{ij}^{(t)} \neq s_{ji}^{(t)}$, $i \neq j$, $i, j = 1, \cdots, n$ for the same n objects and the same T times. This is a collection of telephone communication data. We show an example when $t = 1$. The data is observed as the number of telephone calls from one prefecture to another for 47 prefectures in Japan [70]. For example, the number of telephone calls from A prefecture to B prefecture is different from the number of telephone calls from B prefecture to A prefecture, so we can treat this data as asymmetric similarity data. For the same prefectures, we obtain the data of the number of telephone services sold. The connection between the prefectures consists of four leased circuit services (variables), 0.3kHz - 3.4kHz frequencies band service, voice transmission band service, code service (50bit/s) which is data transmission only, and code service (9600bit/s).

Using the asymmetric similarity data, we create the 94×94 super matrix in equation (3.29). We use the arithmetic mean for the average function. In order to account for the difference between the populations of the prefectures, we normalize the super matrix data with respect to each prefecture. Using model (3.30), we obtain the result of fuzzy clustering, $\hat{u}_{i^{(c)}d_k}$, where the number of clusters is assumed to be 2.

In order to obtain an interpretation for the two clusters obtained, we use equations (3.37) and (4.12) and the data, $x_{i^{(c)}a}$. The latter is the number of telephone services sold with respect to the four variables.

Figure 4.12 shows the result of fuzzy cluster loadings in equation (3.37) for the upper part and the lower part. Figure 4.13 shows the result of kernel fuzzy cluster loading in equation (4.12). We use the polynomial kernel in equation (4.3) where $d = 2$. We obtain two clusters, C_1 and C_2.

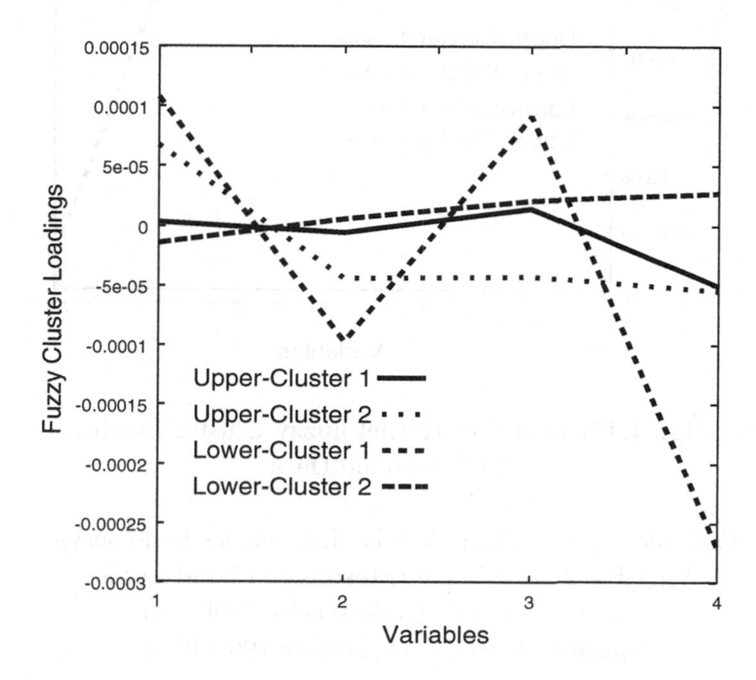

Fig. 4.12. Result of Fuzzy Cluster Loading in Telephone Data

From the results shown in figures 4.12 and 4.13, we can see a significant property of cluster C_1 for receiving part with respect to variables v_3 and v_4, that is code service (50bit/s) and code service (9600bit/s). The reason seems to be that the prefectures in cluster C_1 for lower part received a large number of telephone calls from the central area of Japan. As for the comparison between the two results of figures 4.12 and 4.13, we can see a clearer result in figure 4.13. That is, we have a large difference of cluster C_1 for the upper and lower parts. Otherwise there is no significant difference in cluster C_2 for both the upper and lower parts. The kernel fuzzy cluster loading may be able to avoid the noise when compared with fuzzy cluster loading.

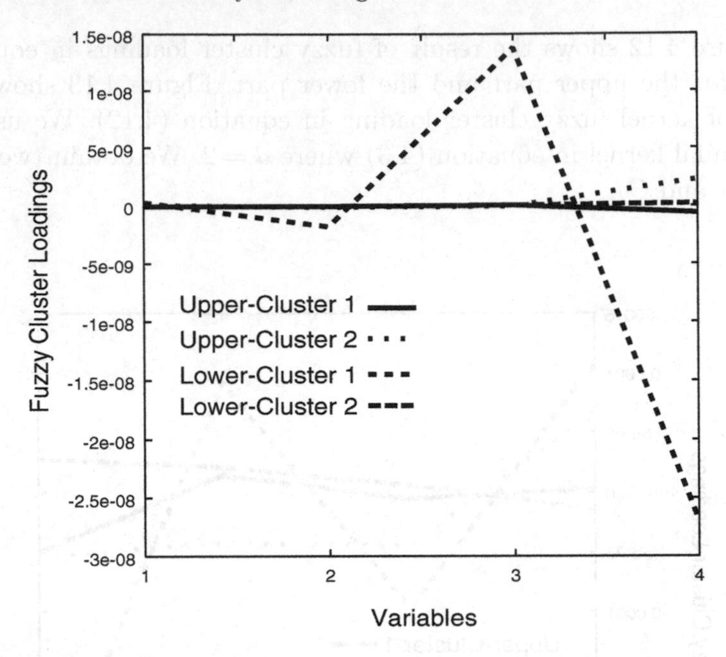

Fig. 4.13. Result of Kernel Fuzzy Cluster Loading in Telephone Data

Variable 1 (v_1): 0.3kHz-3.4kHz frequencies band service
Variable 2 (v_2): Voice transmission band service
Variable 3 (v_3): Code service (50bit/s)
Variable 4 (v_4): Code service (9600bit/s)

5. Evaluation of Fuzzy Clustering

What is natural classification? The search for an answer to this question is the basis of fuzzy clustering. The essence of fuzzy clustering is to consider not only the belonging status of an object to the assumed clusters, but also to consider how much each of the objects belong to the clusters. Fuzzy clustering is a method to obtain "natural groups" in the given observations by using an assumption of a fuzzy subset on clusters. By this method, we can get the degree of belongingness of an object to a cluster. That is, each object can belong to several clusters with several degrees, and the boundaries of the clusters become uncertain. Fuzzy clustering is a natural classification when considering real data. There is no doubt concerning the usefulness of this clustering seeing its wide application to many fields. However, these methods all suffer in that it is difficult to interpret the clusters obtained. Such clustering sometimes causes confusion when trying to understand clustering behavior because each cluster is not exclusive. Each cluster has its own degree of mixing with other clusters. Sometimes, a very mixed cluster is created which makes it difficult to interpret the result. In this case, we need to evaluate the cluster homogeneity in the sense of the degree of belongingness. In this chapter, we describe evaluation techniques [40], [42] for the result of fuzzy clustering using homogeneity analysis [16].

While in the case of hard clustering, the evaluation is simply based on the observations which belong to each of the clusters. However, in the case of fuzzy clustering, we have to consider two concepts. These are the observations and the degree of belongingness which each object possess. We describe a method to obtain the exact differences of the clusters using the idea of homogeneity analysis. The result will show the similarities of the clusters, and will give an indication for the evaluation of a variable number of clusters in fuzzy clustering.

Mika Sato-Ilic and Lakhmi C. Jain: *Innovations in Fuzzy Clustering*, StudFuzz **205**, 105–123 (2006)
www.springerlink.com

The evaluation of the clustering result has two aspects. The first of them is the validity of the clustering result which discusses whether the clustering result is satisfactory, good, or bad. The other is the interpretation of the clusters obtained and a discussion of how it is possible to interpret them. This chapter focuses on the first point. For the interpretation of the fuzzy clustering result, we discuss fuzzy cluster loadings in chapters 3 and 4. The validity of the fuzzy clustering result has also been discussed. In the conventional measures of the validity of fuzzy clustering, partition coefficient and entropy coefficient are well known [5]. These measures are essentially based on the idea that a clear classification is a better result. Using the idea of within-class dispersion and between-class dispersion, separation coefficients are introduced [20]. According to the fuzzy hypervolume, the partition density was discussed [15]. In this chapter, we describe methods of evaluation of a fuzzy clustering result which use the idea of the homogeneity of homogeneity analysis.

5.1 Homogeneity Analysis

Homogeneity analysis is a well known technique for optimizing the homogeneity of variables by manipulation and simplification. Historically, the idea of homogeneity is closely related to the idea that different variables may measure 'the same thing'. We can reduce the number of variables, or put a lower weight for such a variable in order to get fair situations when comparing other variables.

Suppose X and Y are multivariable data matrixes which are $n \times p$, $n \times q$ respectively. Where, n is the number of objects and p and q are the numbers of variables for X and Y, respectively. We assume the weight vectors for X and Y and denote them as $\boldsymbol{a}$ and $\boldsymbol{b}$ which are $p \times 1$ and $q \times 1$ vectors.

The purpose of the homogeneity analysis is to find $\boldsymbol{a}$ and $\boldsymbol{b}$, and the $n \times 1$ vector $\boldsymbol{f}$ to minimize the following:

$$S(\boldsymbol{f}, \boldsymbol{a}, \boldsymbol{b}) = \|\boldsymbol{f} - X\boldsymbol{a}\|^2 + \|\boldsymbol{f} - Y\boldsymbol{b}\|^2. \tag{5.1}$$

For fixed $\boldsymbol{f}$, the estimates of $\boldsymbol{a}$ and $\boldsymbol{b}$ which minimize equation (5.1) are obtained as follows:

$$\boldsymbol{a} = (X^{'}X)^{-1}X^{'}\boldsymbol{f}, \quad \boldsymbol{b} = (Y^{'}Y)^{-1}Y^{'}\boldsymbol{f}.$$

Using the following inequality,

$$\begin{aligned} S(\boldsymbol{f}, \boldsymbol{a}, \boldsymbol{b}) &\geq \|\boldsymbol{f} - P_X \boldsymbol{f}\|^2 + \|\boldsymbol{f} - P_Y \boldsymbol{f}\|^2 \\ &= 2\|\boldsymbol{f}\|^2 - \boldsymbol{f}'(P_X + P_Y)\boldsymbol{f}, \end{aligned} \tag{5.2}$$

the problem of minimizing equation (5.1) is equivalent to that of minimizing equation (5.2). Where, P_X and P_Y are orthogonal projection matrixes of X and Y and defined as:

$$P_X = X(X'X)^{-}X', \quad P_Y = Y(Y'Y)^{-}Y', \tag{5.3}$$

here A^{-} shows the Moore & Penrose inverse matrix of A. We can rewrite equation (5.2) as:

$$S^*(\boldsymbol{f}) = 2\|\boldsymbol{f}\|^2 - \boldsymbol{f}'(P_X + P_Y)\boldsymbol{f}. \tag{5.4}$$

Minimizing equation (5.4) is equivalent to maximizing

$$\boldsymbol{f}'(P_X + P_Y)\boldsymbol{f},$$

under the condition, $\boldsymbol{f}'\boldsymbol{f} = 1$. Therefore, finding $\boldsymbol{f}$ which minimizes equation (5.1) is a problem of finding the eigen-vector of $P_X + P_Y$ corresponding to the maximum eigen-value of $P_X + P_Y$.

5.2 Homogeneity Measure for Symmetric Fuzzy Clustering Result

In order to simplify the status of a cluster, we define the feature vector of cluster k as:

$$X_k \boldsymbol{a}_k = \begin{pmatrix} x_{1_k 1} & x_{2_k 1} & \cdots & x_{n_k 1} \\ x_{1_k 2} & x_{2_k 2} & \cdots & x_{n_k 2} \\ \vdots & \vdots & \vdots & \vdots \\ x_{1_k p} & x_{2_k p} & \cdots & x_{n_k p} \end{pmatrix} \begin{pmatrix} a_{1_k} \\ a_{2_k} \\ \vdots \\ a_{n_k} \end{pmatrix}. \tag{5.5}$$

Where, $\boldsymbol{a}_k$ shows a $n_k \times 1$ vector whose components show the degree of belongingness of objects to a cluster k. X_k is a part of the given data corresponding to the objects of $\boldsymbol{a}_k$.

In chapter 1, we can obtain the result of fuzzy clustering which is shown as the degree of belongingness of objects to clusters. We denote this as $\hat{U} = (\hat{u}_{ik})$, where $i = 1, \cdots, n,\ \ k = 1, \cdots, K$, when the number of objects is n and the number of clusters is K. In this matrix $\hat{U}$, we take the high values of degree ($\geq r$) with respect to each cluster, and make the K vectors,

$$\boldsymbol{a}_1, \cdots, \boldsymbol{a}_K.$$

Here

$$\boldsymbol{a}_k = \begin{pmatrix} a_{1_k} \\ \vdots \\ a_{n_k} \end{pmatrix}, \quad a_l \geq r, \quad k = 1, \cdots, K, \quad l = 1_k, \cdots, n_k.$$

All of the $a_{1_k}, \cdots, a_{n_k}$ are the degree of belongingness for a cluster k, and r is the given value to determine the high contribute to the clusters, and $\frac{1}{K} \leq r \leq 1$. Also, n_k is the number of degrees of belongingness for a cluster k greater than r.

According to the vectors $\boldsymbol{a}_1, \cdots, \boldsymbol{a}_K$, we construct the matrixes using parts of the given data corresponding to the objects which consist of each vector. We denote these matrixes as $X_1, \cdots X_K$, where X_k, $(k = 1, \cdots, K)$ is a $p \times n_k$ matrix where the number of variables is p.

Equation (5.5) means that we extract much higher values of grades to a cluster k and corresponding object's vectors from the data matrix.

The column vectors of X_k are shown as:

$$\boldsymbol{x}_{1_k}, \boldsymbol{x}_{2_k}, \cdots, \boldsymbol{x}_{n_k},$$

where

$$\boldsymbol{x}_{i_k} = \begin{pmatrix} x_{i_k 1} \\ x_{i_k 2} \\ \vdots \\ x_{i_k p} \end{pmatrix},$$

equation (5.5) is then

$$\begin{aligned} X_k \boldsymbol{a}_k &= (\boldsymbol{x}_{1_k}, \boldsymbol{x}_{2_k}, \cdots, \boldsymbol{x}_{n_k}) \begin{pmatrix} a_{1_k} \\ a_{2_k} \\ \vdots \\ a_{n_k} \end{pmatrix} \\ &= \boldsymbol{x}_{1_k} a_{1_k} + \boldsymbol{x}_{2_k} a_{2_k} + \cdots + \boldsymbol{x}_{n_k} a_{n_k} \\ &= \sum_{i_k=1}^{n_k} \boldsymbol{x}_{i_k} a_{i_k}. \end{aligned}$$

Here $\boldsymbol{x}_{i_k}$ shows i-th object and a_{i_k} shows the degree of belongingness of an object i to a cluster k where $a_{i_k} \geq r$. If a cluster k has a property which characterizes the cluster clearly, then all of $a_{1_k}, a_{2_k}, \cdots, a_{n_k}$ are close to 1, and $\boldsymbol{x}_{1_k}, \boldsymbol{x}_{2_k}, \cdots, \boldsymbol{x}_{n_k}$ are similar to each other. So, $X_k\boldsymbol{a}_k$ shows the property of a cluster k in the sense of the degree of belongingness of the objects to a cluster k. Then, by comparing $X_k\boldsymbol{a}_k$ and $X_l\boldsymbol{a}_l$, we can know the similarity between clusters C_k and C_l, in the sense of homogeneity of clusters C_k and C_l.

Then we propose an evaluation function of homogeneity between clusters as follows:

$$H(C_k, C_l) = \|\boldsymbol{f} - X_k\boldsymbol{a}_k\|^2 + \|\boldsymbol{f} - X_l\boldsymbol{a}_l\|^2. \tag{5.6}$$

Where $\boldsymbol{a}_k$ and $\boldsymbol{a}_l$ are the fuzzy grades of clusters C_k and C_l respectively. These are $\boldsymbol{a}_k,\ \boldsymbol{a}_l \geq r$. $H(C_k, C_l)$ shows the dissimilarity of clusters C_k and C_l based on the meaning of homogeneity. $\boldsymbol{f}$ is a $p \times 1$ vector and satisfies $\boldsymbol{f}'\boldsymbol{f} = 1$.

The algorithm used to find the measure of homogeneity between a pair of clusters is:

(Step 1) Using $X_k,\ X_l,\ \ k,l = 1, \cdots, K,\ \ k \neq l$, calculate P_{X_k} and P_{X_l} from equation (5.3).

(Step 2) From $P_{X_k} + P_{X_l}$, calculate the maximum eigen-value λ and its corresponding eigen-vector $\boldsymbol{f}$.

(Step 3) Calculate the measure of homogeneity for clusters C_k and C_l using equation (5.6).

5.3 Homogeneity Measure for an Asymmetric Fuzzy Clustering Result

This section presents a clustering method that captures the transition of asymmetric structures in 3-way asymmetric similarity data. Usually 3-way asymmetric similarity data forms asymmetric similarity data for several situations. The asymmetric property is represented by the asymmetry of the clusters, and are common for all situations. This measure captures the transition structure for the movement of asymmetry for changing situations. We introduce a measure that shows the homogeneity status of the asymmetric structures.

The relation between objects is represented in terms of proximity, similarity (or dissimilarity), and connection. In this section we use similarity, especially for asymmetric similarity data. For example, the degree of liking or disliking in human relationships, or the degree of confusion based on perception for a discrimination problem. Inflow and outflow of materials or information between several areas that are represented in industry relationships by qualities such as mobility data. For asymmetric data, asymmetry itself has an essential meaning. That is why techniques based on asymmetric similarity have generated great interest among a number of researchers.

In conventional clustering methods for asymmetric similarity data, A. D. Gordon [17] proposed a method using only the symmetric part of the data. This method is based on the idea that the asymmetry of the given similarity data can be regarded as errors of the symmetric similarity data. L. Hubert [24] proposed a method to select a maximum element in the corresponding elements of the similarity matrix. What these methods do is to essentially change the asymmetric data to symmetric data. However, a need exists to analyze the asymmetric data directly. H. Fujiwara [14] has extended Hubert's method and proposed a technique that can determine the similarity between clusters while considering asymmetry in the area of hierarchical clustering methods.

In this section, we discuss a case in which an asymmetric data was observed for several times or situations. Then we propose a method that can capture one fuzzy clustering result, as well as, several asymmetric similarity results between clusters for the several situations. This model seems to be similar to DEDICOM for 3-way data [28]. In the model the condition of the solution is different, due to the fuzzy clustering condition, and therefore, the domain of the solution is different. Note that the clusters obtained through times will not be changed, so we can compare the result of each asymmetric similarity between clusters for each situation. Due to this comparison of the asymmetric structure, we need a concept to determine the difference of the asymmetric structure. This problem is caused by an inter-relation among objects, so it can not be measured by the normal Euclidean norm of the matrixes, which can show the asymmetric similarity between clusters. We introduce the homogeneity measure to evaluate the difference of several asymmetric structures of common clusters.

5.3.1 Asymmetric Fuzzy Clustering Model

The asymmetric fuzzy clustering model [36] is as follows:

$$s_{ij} = \sum_{k=1}^{K} \sum_{l=1}^{K} w_{kl} u_{ik} u_{jl} + \varepsilon_{ij}, \tag{5.7}$$

where s_{ij} is the observed asymmetric similarity between objects i and j, and $0 \leq s_{ij} \leq 1$. s_{ij} does not equal s_{ji} when $i \neq j$. u_{ik} is a fuzzy grade which represents the degree of belongingness of an object i to a cluster k. Generally, u_{ik} are denoted by using the matrix representation $U = (u_{ik})$ called a partition matrix, which satisfy the condition (1.1). In this case, the product $u_{ik} u_{jk}$ is the degree of simultaneous belongingness of objects i and j to a cluster k. That is, the product denotes the degree of sharing of common properties. In this model, the weight w_{kl} is considered to be a quantity which shows the asymmetric similarity between a pair of clusters. That is, we assume that the asymmetry of the similarity between the objects is caused by the asymmetry of the similarity between the clusters.

5.3.2 Generalized Asymmetric Fuzzy Clustering Model for 3-Way Data

If the observed data was obtained as a $n \times n$ asymmetric similarity matrix of objects and such a data was obtained for several situations, then the model which is shown in equation (5.7) can not be used. As a result, we extended the model for the 3-way data as follows:

$$s_{ij}^{(t)} = \sum_{k=1}^{K} \sum_{l=1}^{K} w_{kl}^{(t)} u_{ik} u_{jl} + \varepsilon_{ij}. \tag{5.8}$$

Where, $s_{ij}^{(t)}$ shows the similarity between i-th and j-th objects at t-th situation and $1 \leq t \leq T$. And $w_{kl}^{(t)}$ is the similarity between k and l clusters at t-th situation and $w_{kl}^{(t)} \neq w_{lk}^{(t)}$, $k \neq l$. Note that we can obtain a common solution through situations as $\hat{U} = (\hat{u}_{ik})$ but different asymmetric relations of clusters for each situation as $\hat{W}^{(t)} = (\hat{w}_{kl}^{(t)})$. The clusters obtained for all situations will not change, so we could compare the asymmetric results $\hat{W}^{(1)}, \hat{W}^{(2)}, \cdots, \hat{W}^{(T)}$ mathematically.

5.3.3 Comparison of the Structure of Asymmetry

We now discuss how to evaluate the differences of the asymmetric structure. An asymmetric structure depends on neither a pair of asymmetric relations between objects nor the distance of all of elements of the asymmetric matrix. These elements are able to show the asymmetric relationship among the clusters. It may however depend on the inter-relationship between the objects. We introduce the concept of homogeneity, based on the homogeneity analysis mentioned in section 5.1.

In general, if we observed m data matrixes as $X_1, \cdots, X_m$, we can then obtain the solution $\boldsymbol{f}$ from $P_{X_1} + \cdots + P_{X_m}$ as discussed in section 5.1. If we assume the weights for the matrixes $X_1, \cdots, X_m$, are $\boldsymbol{a}_1, \cdots, \boldsymbol{a}_m$ respectively, then the estimates of the weights $\boldsymbol{a}_i, i = 1, \cdots, m$, which minimize

$$S(\boldsymbol{f}, \boldsymbol{a}_i \mid i = 1, \cdots, m) = \sum_{i=1}^{m} \|\boldsymbol{f} - X_i \boldsymbol{a}_i\|,$$

can be obtained as follows:

$$\boldsymbol{a}_i = (X_i' X_i)^{-1} X_i' \boldsymbol{f}. \tag{5.9}$$

Using the idea of homogeneity we compare the asymmetric similarity which can be obtained for the several situations $W^{(1)}, W^{(2)}, \cdots, W^{(T)}$. Suppose we obtain estimates $\hat{W}^{(1)}, \hat{W}^{(2)}, \cdots, \hat{W}^{(T)}$ by using model (5.8). From

$$P_{\hat{W}^{(1)}} + \cdots + P_{\hat{W}^{(T)}},$$

we can obtain the $K \times 1$ vector $\boldsymbol{f}$ as the eigen-vector corresponding to the maximum eigen-value of $P_{\hat{W}^{(1)}} + \cdots + P_{\hat{W}^{(T)}}$. Here,

$$P_{\hat{W}^{(t)}} = \hat{W}^{(t)} (\hat{W}^{(t)'} \hat{W}^{(t)})^{-} \hat{W}^{(t)'}, \ \ t = 1, \cdots, T.$$

From equation (5.9), we find $\boldsymbol{a}_1, \cdots, \boldsymbol{a}_T$, respectively. Here,

$$\boldsymbol{a}_t = (\hat{W}^{(t)'} \hat{W}^{(t)})^{-1} \hat{W}^{(t)'} \boldsymbol{f}, \ \ t = 1, \cdots, T. \tag{5.10}$$

The evaluation measure on homogeneity for the asymmetric structures $\hat{W}^{(t)}$ and $\hat{W}^{(s)}$ $(t, s = 1, \cdots, T)$, is defined as:

$$e(\hat{W}^{(t)}, \hat{W}^{(s)}) = (\boldsymbol{a}_t - \boldsymbol{a}_s)' (\boldsymbol{a}_t - \boldsymbol{a}_s). \tag{5.11}$$

This shows how close the weights are to each other. From equations (5.10) and (5.11),

$$
\begin{aligned}
(\boldsymbol{a}_t - \boldsymbol{a}_s)'(\boldsymbol{a}_t - \boldsymbol{a}_s) = \boldsymbol{f}'[&\hat{W}^{(t)}\{(\hat{W}^{(t)'}\hat{W}^{(t)})^2\}^{-1}\hat{W}^{(t)'} \\
&-2\hat{W}^{(t)}(\hat{W}^{(t)'}\hat{W}^{(t)})^{-1}(\hat{W}^{(s)'}\hat{W}^{(s)})^{-1}\hat{W}^{(s)'} \\
&+\hat{W}^{(s)}\{(\hat{W}^{(s)'}\hat{W}^{(s)})^2\}^{-1}\hat{W}^{(s)'}]\boldsymbol{f}.
\end{aligned}
\tag{5.12}
$$

The value of equation (5.12) is small, if

$$
\boldsymbol{f}'(\hat{W}^{(t)}(\hat{W}^{(t)'}\hat{W}^{(t)})^{-1}(\hat{W}^{(s)'}\hat{W}^{(s)})^{-1}\hat{W}^{(s)'})\boldsymbol{f}
$$

is large. Also, $\boldsymbol{f}$ is given as the result of the homogeneity analysis. Essentially, equation (5.11) evaluates $\hat{W}^{(t)}(\hat{W}^{(t)'}\hat{W}^{(t)})^{-1}(\hat{W}^{(s)'}\hat{W}^{(s)})^{-1}$ $\hat{W}^{(s)'}$.

5.4 Numerical Examples for the Evaluation of the Fuzzy Clustering Result

5.4.1 Numerical Examples for Evaluation of Symmetric Fuzzy Clustering Result

The data is Landsat data which was observed over the Kushiro marsh-land in May, 1992. The value of the data shows the amount of light reflected from the ground for six kinds of light. Figure 3.8 shows the result obtained by the use of the fuzzy c-means method where the control parameter of fuzziness $m = 1.25$ in equation (1.3).

As we described in section 3.3.2, the data shows the area of water, the area of mountains and the area of the cities. This means that it is necessary to consider three kinds of clusters. In the figure 3.8 each of the three axes represents one of the sets of clusters. The symbols used to denote mountainous areas, areas of water and areas of city are $\times$, $+$, $*$ respectively. The value of the coordinate for each of the clusters is the degree of belongingness for each cluster. From the centroids for each cluster and the data, we can see that cluster C_1 shows the mountain area, cluster C_2 indicates water, that is marsh-land or river, and cluster C_3 is city area. Using equation (5.6) we obtain the result of the measure of homogeneity for a pair of clusters as table 5.1.

Table 5.1 Measure of Homogeneity of the Clusters

Clusters	C_1	C_2	C_3
C_1	*	31668011	29195197
C_2	31668011	*	44305525
C_3	29195197	44305525	*

From this result, we can see that the value of the homogeneity between cluster C_1 and cluster C_3 shows that the mountain area and the city area are more similar than the relation between cluster C_2 and cluster C_3. We can see that this is different from the simple overlapping status of the degree of belongingness shown in Figure 3.8.

The second example used the data for 50 AIDS patients for two times with respect to the main variables for HIV. These are called CD4+ and CD8+. Here, the two extended relational fuzzy c-means methods for 3-way data [41] discussed in chapter 1 are used. One method uses the idea of a multicriteria optimization problem to obtain the solution as a pareto optimum clustering. Using this method, we get one clustering result in a number of situations. The other result uses the idea of dynamic clustering [38] to see dynamic changes in the observations.

The fuzzy clustering results of equations (1.6) and (1.8) are shown in figures 5.1 and 5.2. In these figures, the abscissa shows the number of patients and the ordinate shows the degree of belongingness of the patients to cluster C_1. In this case, the number of clusters is given as 2. Thus, it is enough to show the degree to cluster C_1, due to the condition that the sum of the degree of belongingness of objects with respect to clusters is 1. In particular, the dotted lines in these figures show the lines of the degree as 0.5, that is, the upper dots of these lines show the patients who almost belong to cluster C_1, and the lower dots show the patients who almost belong to cluster C_2.

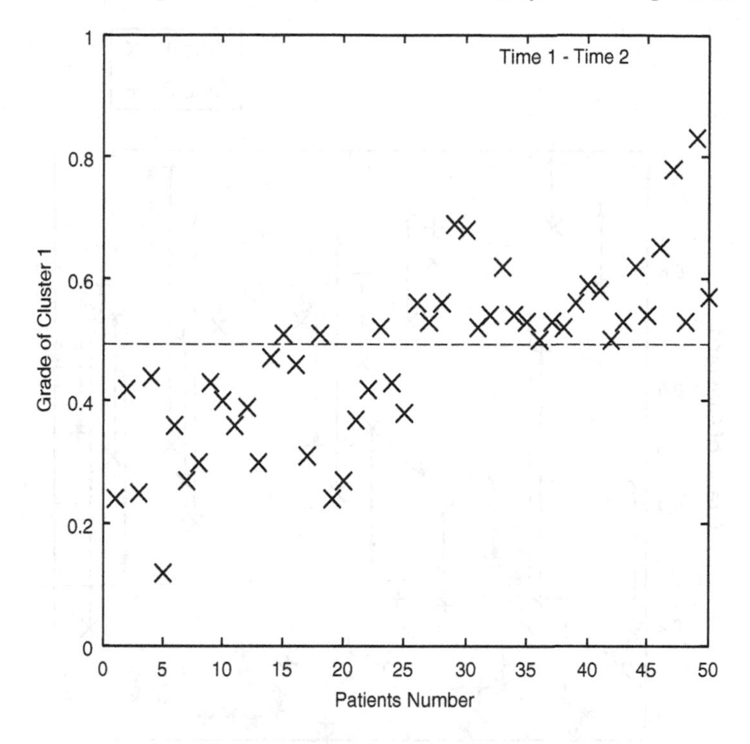

Fig. 5.1. Result of Equation (1.6)

In equation (1.6), we use the same weight for the first and the second time. That is, we use $w^{(t)} = 0.5, \quad t = 1, 2$ in equation (1.6). From figure 5.1, we see the two clusters obtained and the classification situation over the two times. In this method, we obtain a clustering result which involves the inherent changing situation from the first to the second time. We obtain the clusters in which patients are similar to each other and patterns of the change of patients with respect to the variables from the first to the second time are similar to each other.

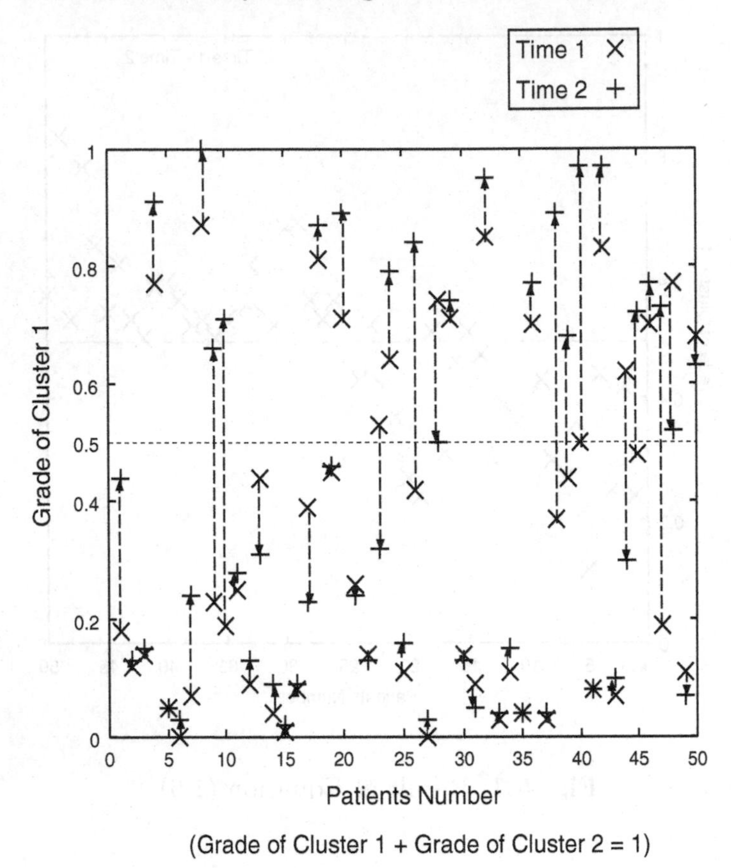

Fig. 5.2. Result of Equation (1.8)

In figure 5.2, the symbol '×' shows the degree of belongingness to the cluster C_1 at the first time and the symbol '+' shows at the second time. The arrow shows the movement from the first time to the second time. From this figure, we can see the remarkable movement of the lower area at the first time to the upper area at the second time. That is, patients moved from cluster C_2 to cluster C_1 due to the movement from the first time point to the second. According to the data, cluster C_2 shows the patients who have a high number of CD8+, which is a well known indication of the illness' progress, have a decrease in the number of CD8+. So, it is thought that these patient's illness might proceed according to the time.

Table 5.2 shows the resultant homogeneity of the clusters and a comparison of the two methods shown in equations (1.6) and (1.8). In this table, H_1, H_2, H_3, and H_4 are as follows:

$$H_1 = \|\boldsymbol{f} - X_1\boldsymbol{a}_1\|^2 + \|\boldsymbol{f} - X_2\boldsymbol{a}_2\|^2,$$
$$H_2 = \|\boldsymbol{f} - X_1\boldsymbol{a}_3\|^2 + \|\boldsymbol{f} - X_2\boldsymbol{a}_3\|^2,$$
$$H_3 = \|\boldsymbol{f} - X_1\boldsymbol{a}_1\|^2 + \|\boldsymbol{f} - X_1\boldsymbol{a}_3\|^2,$$
$$H_4 = \|\boldsymbol{f} - X_2\boldsymbol{a}_2\|^2 + \|\boldsymbol{f} - X_2\boldsymbol{a}_3\|^2.$$

Here X_1 and X_2 are patients data for the first and the second times respectively. $\boldsymbol{a}_1$ shows the degree of belongingness of equation (1.6) at the first time point and $\boldsymbol{a}_2$ is the result at the second time point. $\boldsymbol{a}_3$ shows the degree of belongingness of equation (1.8). From this, we can see that the result of equation (1.8) is similar to the result of equation (1.6) at the first time in the sense of homogeneity. In other words, the pareto optimum clustering result shown in equation (1.6) seems to capture the salience of the first time rather than at the second time.

Table 5.2 Results of Homogeneity of Clusters

H_1	H_2	H_3	H_4
14.9×10^8	14.5×10^8	13.8×10^8	15.5×10^8

5.4.2 Numerical Examples for Evaluation of Asymmetric Fuzzy Clustering Result

The observed data is shown in table 5.3 [34]. This data shows the human relations between 16 children. The value of this data shows the degree of liking and disliking between the children. The number of clusters is determined as 5 and is based on the value of fitness.

Table 5.4 shows the result of the degree of belongingness of the children to the clusters using model (5.7). In this table, C_1, C_2, C_3, C_4, and C_5 represent the five clusters. The children who are in the same cluster are good friends with each other. C_1 and C_2 are clusters which consist of boys only, and C_3, C_4, C_5 are girls only.

Table 5.5 shows the values of $\hat{w}_{kl}$, which is the relationship between the clusters. For example, the boys who belong to cluster C_2 have an interest in the girls who belong to cluster C_5, because the value of $\hat{w}_{25}$ is large. The girls of C_5 are not interested in the boys of C_2, because the value of $\hat{w}_{52}$ is small.

Table 5.6 shows the data where the values for the children in cluster C_2 reverse positions with the children in cluster C_5. Table 5.7 shows similar data for clusters C_3 and C_4. In tables 5.6 and 5.7, the boldfaced values indicate the changed values from the original data in table 5.3. For the data shown in tables 5.3, 5.6, and 5.7, we created an artificial data. Table 5.3 is data for the first situation. Table 5.6 is the second, and table 5.7 is the third situation. Using this 3-way data, we use the proposed model shown in equation (5.8). To get the solution the initial values are random numbers and the result is shown as tables 5.8 and 5.9. The fitness of the data and the model is 0.01.

Table 5.3. Dissimilarity Matrix for the Children

Child	1	2	3	4	5	6	7	8	9	10	11	12	13	14	15	16
1	0	2	3	3	1	1	2	1	3	6	2	3	6	4	6	4
2	6	0	1	1	6	6	6	6	1	6	6	2	6	2	6	2
3	6	1	0	2	6	6	6	6	1	6	6	1	6	2	6	2
4	6	1	2	0	6	6	6	6	1	6	6	2	6	1	6	2
5	1	3	3	4	0	1	1	2	4	6	3	3	6	4	6	4
6	1	3	2	4	1	0	2	2	3	6	3	2	6	3	6	3
7	1	3	3	4	1	1	0	2	4	6	3	3	6	4	6	4
8	6	1	2	2	6	6	6	0	2	6	1	3	6	3	6	3
9	6	6	6	6	6	6	6	6	0	6	6	1	6	1	6	1
10	6	2	3	3	6	6	6	2	3	0	1	4	1	4	1	4
11	6	1	2	2	6	6	6	1	2	6	0	3	6	3	6	3
12	6	6	6	6	6	6	6	6	1	6	6	0	6	1	6	1
13	6	2	3	3	6	6	6	2	3	1	1	4	0	4	1	4
14	6	6	6	6	6	6	6	6	1	6	6	1	6	0	6	1
15	6	2	3	3	6	6	6	2	3	1	1	4	1	4	0	4
16	6	6	6	6	6	6	6	6	1	6	6	1	6	1	6	0

Table 5.4 Fuzzy Clustering Result for the Data on the Children

Child	C_1	C_2	C_3	C_4	C_5
1	0.94	0.00	0.04	0.02	0.00
2	0.00	0.91	0.09	0.00	0.00
3	0.06	0.87	0.04	0.00	0.03
4	0.00	0.88	0.00	0.08	0.04
5	0.98	0.00	0.00	0.02	0.00
6	0.94	0.04	0.00	0.01	0.01
7	0.95	0.00	0.00	0.05	0.00
8	0.06	0.00	0.94	0.00	0.00
9	0.00	0.06	0.00	0.00	0.94
10	0.00	0.00	0.00	1.00	0.00
11	0.00	0.00	0.91	0.09	0.00
12	0.05	0.00	0.00	0.00	0.95
13	0.00	0.00	0.00	1.00	0.00
14	0.00	0.00	0.00	0.03	0.97
15	0.00	0.00	0.00	1.00	0.00
16	0.02	0.00	0.00	0.03	0.95

Table 5.5 Similarity between Clusters

Clusters	C_1	C_2	C_3	C_4	C_5
C_1	0.92	0.47	0.66	0.00	0.42
C_2	0.01	0.98	0.02	0.01	0.77
C_3	0.00	0.76	0.98	0.00	0.56
C_4	0.00	0.56	0.77	0.88	0.37
C_5	0.01	0.00	0.00	0.01	0.92

Table 5.6 Dissimilarity matrix of Children which Changed the Positions of Clusters C_2 and C_5

Child	1	2	3	4	5	6	7	8	9	10	11	12	13	14	15	16
1	0	2	3	3	1	1	2	1	3	6	2	3	6	4	6	4
2	6	0	1	1	6	6	6	6	**6**	6	6	**6**	6	**6**	6	**6**
3	6	1	0	2	6	6	6	6	**6**	6	6	**6**	6	**6**	6	**6**
4	6	1	2	0	6	6	6	6	**6**	6	6	**6**	6	**6**	6	**6**
5	1	3	3	4	0	1	1	2	4	6	3	3	6	4	6	4
6	1	3	2	4	1	0	2	2	3	6	3	2	6	3	6	3
7	1	3	3	4	1	1	0	2	4	6	3	3	6	4	6	4
8	6	1	2	2	6	6	6	0	2	6	1	3	6	3	6	3
9	6	**1**	**1**	**1**	6	6	6	6	0	6	6	1	6	1	6	1
10	6	2	3	3	6	6	6	2	3	0	1	4	1	4	1	4
11	6	1	2	2	6	6	6	1	2	6	0	3	6	3	6	3
12	6	**2**	**1**	**2**	6	6	6	6	1	6	6	0	6	1	6	1
13	6	2	3	3	6	6	6	2	3	1	1	4	0	4	1	4
14	6	**2**	**2**	**1**	6	6	6	6	1	6	6	1	6	0	6	1
15	6	2	3	3	6	6	6	2	3	1	1	4	1	4	0	4
16	6	**2**	**2**	**2**	6	6	6	6	1	6	6	1	6	1	6	0

Table 5.7 Dissimilarity matrix of Children which Changed the Positions of Clusters C_3 and C_4

Child	1	2	3	4	5	6	7	8	9	10	11	12	13	14	15	16
1	0	2	3	3	1	1	2	1	3	6	2	3	6	4	6	4
2	6	0	1	1	6	6	6	6	1	6	6	2	6	2	6	2
3	6	1	0	2	6	6	6	6	1	6	6	1	6	2	6	2
4	6	1	2	0	6	6	6	6	1	6	6	2	6	1	6	2
5	1	3	3	4	0	1	1	2	4	6	3	3	6	4	6	4
6	1	3	2	4	1	0	2	2	3	6	3	2	6	3	6	3
7	1	3	3	4	1	1	0	2	4	6	3	3	6	4	6	4
8	6	1	2	2	6	6	6	0	2	**2**	1	3	**2**	3	**2**	3
9	6	6	6	6	6	6	6	6	0	6	6	1	6	1	6	1
10	6	2	3	3	6	6	6	**6**	3	0	**6**	4	1	4	1	4
11	6	1	2	2	6	6	6	1	2	**1**	0	3	**1**	3	**1**	3
12	6	6	6	6	6	6	6	6	1	6	6	0	6	1	6	1
13	6	2	3	3	6	6	6	**6**	3	1	**6**	4	0	4	1	4
14	6	6	6	6	6	6	6	6	1	6	6	1	6	0	6	1
15	6	2	3	3	6	6	6	**6**	3	1	**6**	4	1	4	0	4
16	6	6	6	6	6	6	6	6	1	6	6	1	6	1	6	0

From table 5.8, we see that in this case an almost identical result as that shown in table 5.4 was obtained. According to the similarity among clusters for each situation (shown in table 5.9), we can see clearly the difference of the asymmetric structures through situations. That is, in table 5.9, in the second situation, the values of $\hat{w}_{25}$ and $\hat{w}_{52}$ (shown as boldface in table 5.9 (second situation)) have clearly reversed position to the first situation. At the third situation the asymmetric features of clusters C_3 and C_4 are changed when compared to the first situation. This can be seen in the values of $\hat{w}_{34}$ and $\hat{w}_{43}$. This is shown as boldface in table 5.9 (third situation) at the first and third situations. This shows the validity of the method.

Table 5.8 Fuzzy Clustering Result of Children Data for Several Situations

Child	C_1	C_2	C_3	C_4	C_5
1	1.00	0.00	0.00	0.00	0.00
2	0.00	0.99	0.00	0.00	0.00
3	0.04	0.93	0.00	0.00	0.01
4	0.00	0.92	0.00	0.04	0.03
5	1.00	0.00	0.00	0.00	0.00
6	0.97	0.00	0.00	0.00	0.02
7	1.00	0.00	0.00	0.00	0.00
8	0.04	0.00	0.95	0.00	0.00
9	0.00	0.00	0.00	0.00	1.00
10	0.00	0.00	0.00	1.00	0.00
11	0.00	0.00	0.94	0.05	0.00
12	0.02	0.00	0.00	0.00	0.97
13	0.00	0.00	0.00	1.00	0.00
14	0.00	0.00	0.00	0.00	1.00
15	0.00	0.00	0.00	1.00	0.00
16	0.00	0.00	0.00	0.01	0.98

Table 5.9 Similarity between Clusters for Several Situations

(First Situation)

Clusters	C_1	C_2	C_3	C_4	C_5
C_1	0.98	0.59	0.77	0.00	0.50
C_2	0.00	0.99	0.00	0.00	**0.90**
C_3	0.00	0.91	0.99	**0.00**	0.66
C_4	0.00	0.67	**0.91**	0.99	0.45
C_5	0.00	**0.00**	0.00	0.00	0.99

(Second Situation)

Clusters	C_1	C_2	C_3	C_4	C_5
C_1	0.98	0.58	0.77	0.00	0.50
C_2	0.00	0.99	0.00	0.00	**0.00**
C_3	0.00	0.91	0.99	0.00	0.66
C_4	0.00	0.67	0.91	0.99	0.45
C_5	0.00	**0.91**	0.00	0.00	0.99

(Third Situation)

Clusters	C_1	C_2	C_3	C_4	C_5
C_1	0.98	0.59	0.77	0.00	0.50
C_2	0.00	0.99	0.00	0.00	0.90
C_3	0.00	0.89	0.99	**0.91**	0.66
C_4	0.00	0.68	**0.00**	0.99	0.45
C_5	0.00	0.00	0.00	0.00	0.99

We examined the homogeneity measure and the distance of the asymmetric structures for the above situations. The result of the homogeneity using equation (5.11) is shown in table 5.10. Table 5.11 shows the distance for the difference for the three matrixes $\hat{W}^{(1)}$, $\hat{W}^{(2)}$, and $\hat{W}^{(3)}$ which are shown in table 5.9. In this case, $\hat{W}^{(1)}$ is the matrix which is shown in table 5.9 (first situation). $\hat{W}^{(2)}$ is the matrix of table 5.9 (second situation). $\hat{W}^{(3)}$ is the matrix of table 5.9 (third situation). The comparison of tables 5.10 and 5.11 shows that there is a clear difference. In table 5.10, the difference between $\hat{W}^{(1)}$ and $\hat{W}^{(3)}$ is smaller than the difference between $\hat{W}^{(1)}$ and $\hat{W}^{(2)}$. In table 5.11, an opposite result was obtained. This shows that many of the changed objects have a complex connection with each other. That is, in the case of cluster C_5, the children who belong to this cluster do not like children in the other clusters. The children who belong to other clusters are fond of the children in C_5. This shows that the affection is a one way communication. When the situation for this cluster is changed, the global structure is greatly affected.

Table 5.10 Homogeneity of Asymmetric Structures

Asymmetric Matrixes	$\hat{W}^{(1)}$	$\hat{W}^{(2)}$	$\hat{W}^{(3)}$
$\hat{W}^{(1)}$	*	1.712	1.464
$\hat{W}^{(2)}$	1.712	*	1.463
$\hat{W}^{(3)}$	1.464	1.463	*

Table 5.11 Distance of Asymmetric Matrixes

Asymmetric Matrixes	$\hat{W}^{(1)}$	$\hat{W}^{(2)}$	$\hat{W}^{(3)}$
$\hat{W}^{(1)}$	*	0.066	0.067
$\hat{W}^{(2)}$	0.066	*	0.133
$\hat{W}^{(3)}$	0.067	0.133	*

6. Self-Organized Fuzzy Clustering

This chapter presents several methods based on self-organized dissimilarity or similarity [54]. The first uses fuzzy clustering methods while the second is a hybrid method of fuzzy clustering and multidimensional scaling (MDS) [18], [31]. Specifically, a self-organized dissimilarity (or similarity) is defined that uses the result of fuzzy clustering where the dissimilarity (or similarity) of objects is influenced by the dissimilarity (or similarity) of the classification situations corresponding to the objects. In other words, the dissimilarity (or similarity) is defined under an assumption that similar objects have similar classification structures. Through an empirical evaluation the proportion and the fitness of the results of the method, which uses MDS combined with fuzzy clustering, is shown to be effective when using real data. By exploiting the self-organized dissimilarity (or similarity), the defuzzification of fuzzy clustering can cope with the inherent classification structures.

Recently, in the area of data analysis, there is a great interest in the treatment of complicated and huge amounts of data. Hybrid techniques are increasingly popular for such data in a variety of applications. One reason hybrid techniques have drawn attention is that the complexity of the data creates a need to combine the merits of different methods in order to analyze the data. Since each method has a particular merit for data analysis, complex data requires multiplex merits to capture the real feature of the data, due to the multiple aspects of the latent features of complex data. Another reason is the possibility that we can capture new features of the data through the combination of several methods.

For combining different methods of data analysis, there are two practical approaches. The first is to estimate the multiple features of the data simultaneously. Several methods can simultaneously capture the classification feature and the regression feature of the data [23]. Several multivariate data analyses have been proposed which use fuzzy

Mika Sato-Ilic and Lakhmi C. Jain: *Innovations in Fuzzy Clustering*, StudFuzz **205**, 125–145 (2006)
www.springerlink.com

logic, neural networks, and the self-organized method [55], [62], [65]. We take the approach of using the result of one method directly for a second method. This approach can not be reduced to a simultaneous optimization. The merit is that autonomously clear solutions can be obtained by using the methods separately.

This chapter will focus on the second approach. We obtain the result by fuzzy clustering, and then use this result again for fuzzy clustering in order to obtain a clearer result. In addition we will use the result of fuzzy clustering for multidimensional scaling to propose a new hybrid technique.

In order to present the new techniques, we propose a self-organized dissimilarity (or similarity) using the result of fuzzy clustering. This dissimilarity (or similarity) has a combination of two relevant features the classification structure obtained in the solution space of fuzzy clustering and the distance structure on the metric space in the object's space.

The theories of self-organized techniques are now well-known and many applications have proved their efficiency. The field of data analysis is no exception and much research based on this idea has been proposed. This is especially true for, clustering methods which are important for the self organization based data analysis [30], [63].

Given the self-organized dissimilarity, the proposed methods tend to give clearer and crisper results. Although the merit of fuzzy clustering is obtaining natural clustering represented by continuous values of the degree of belongingness, we sometimes have need to reduce the fuzziness of the degree of belongingness in order to obtain a crisp result. The conventional method for removing fuzziness from fuzzy clustering is that an object belongs to a cluster whose degree of belongingness is the largest of the other clusters.

This is inappropriate when an object's highest value of belongingness are almost the same over several clusters. We have to determine the one cluster that the object belongs to with the highest degree. A situation may occur where the degree of belongingness amongst the highest value clusters are almost the same. For example 0.4999 and 0.5001 when there are two clusters. Our proposed method can rectify this situation by exploiting the self-organized dissimilarity (or similarity), while still ensuring the likely return is the highest.

6.1 Fuzzy Clustering Methods based on Self-Organized Similarity

6.1.1 Generalized Fuzzy Clustering Model

The structure of an observed similarity is usually unknown and complicated. Consequently various fuzzy clustering models are required to identify the latent structure of the similarity data. We therefore define a general class of fuzzy clustering models, in order to represent many different structures of similarity data [35], [36]. The merits of the fuzzy clustering models are: (1) the amount of computations necessary for the identification of the models are much fewer than in a hard clustering model, and (2) a suitable fitness can be obtained by using fewer clusters. In the generalized clustering model, aggregation operators are used to define the degree of simultaneous belongingness of a pair of objects to a cluster.

Suppose that there exist K fuzzy clusters on a set of n objects, that is, the partition matrix $U = (u_{ik})$ is given. Let $\rho(u_{ik}, u_{jl})$ be a function which denotes a degree of simultaneous belongingness of a pair of objects i and j to clusters k and l, namely, a degree of sharing common properties. Then a general model for the similarity s_{ij} is defined as follows:

$$s_{ij} = \varphi(\rho_{ij}) + \varepsilon_{ij}, \tag{6.1}$$

where

$$\rho_{ij} = (\rho(u_{i1}, u_{j1}), \cdots, \rho(u_{i1}, u_{jK}), \cdots, \rho(u_{iK}, u_{j1}), \cdots, \rho(u_{iK}, u_{jK})).$$

We assume that if all of $\rho(u_{ik}, u_{jl})$ are multiplied by α, then the similarity is also multiplied by α. Therefore, the function φ itself must satisfy the condition of "positively homogeneous of degree 1 in the ρ", that is,

$$\alpha\varphi(\rho_{ij}) = \varphi(\alpha\rho_{ij}), \quad \alpha > 0.$$

We consider the following function as a typical function of φ:

$$s_{ij} = \{\sum_{k=1}^{K} \rho^r(u_{ik}, u_{jk})\}^{\frac{1}{r}} + \varepsilon_{ij}, \quad 0 < r < +\infty. \tag{6.2}$$

In the following we assume that $r = 1$ in equation (6.2).

The degree ρ is assumed to satisfy the conditions of boundary, monotonicity, and symmetry described in section 3.1.3. In the case of $\rho(u_{ik}, u_{jl}) = u_{ik}u_{jl}$, equation (6.1) is represented as follows:

$$s_{ij} = \sum_{k=1}^{K} u_{ik}u_{jk} + \varepsilon_{ij}. \tag{6.3}$$

Here, u_{ik} is a fuzzy grade which represents the degree of belongingness of an object i to a cluster k, and satisfies the condition shown in equation (1.1). The product $u_{ik}u_{jk}$ is the degree of simultaneous belongingness of objects i and j to a cluster k. That is, the product denotes the degree of the sharing of common properties.

6.1.2 Similarity between Clusters

If the observed similarity data is asymmetric, then the additive fuzzy clustering models in the previous section are not suitable. We extend the model (6.1) as:

$$s_{ij} = \sum_{k=1}^{K}\sum_{l=1}^{K} w_{kl}\rho(u_{ik}, u_{jl}) + \varepsilon_{ij}. \tag{6.4}$$

In this model, the weight w_{kl} is considered to be a quantity which shows the asymmetric similarity between a pair of clusters. That is, we assume that the asymmetry of the similarity between the objects is caused by the asymmetry of the similarity between the clusters.

In the case of $\rho(u_{ik}, u_{jl}) = u_{ik}u_{jl}$, equation (6.4) is represented as equation (5.7). If $k = l$, $u_{ik}, u_{jl} = 1$ and $w_{kl} > 1$, then the right hand side of equation (5.7) clearly exceeds 1.0. Hence we need at least the following condition

$$0 \leq w_{kl} \leq 1, \tag{6.5}$$

because $0 \leq s_{ij} \leq 1$. The method of fuzzy clustering based on this model is to find the partition matrix $U = (u_{ik})$ and $W = (w_{kl})$ which satisfy the conditions (1.1), (6.5) and have the best fitness for model (5.7). Then we find U and W which minimize the following sum of squares error κ^2 under the conditions (1.1) and (6.5),

$$\kappa^2 = \sum_{i \neq j=1}^{n} (s_{ij} - \sum_{k=1}^{K}\sum_{l=1}^{K} w_{kl}u_{ik}u_{jl})^2. \tag{6.6}$$

If $U_{ij} = \sum_{k=1}^{K} \sum_{l=1}^{K} w_{kl} u_{ik} u_{jl}$, then equation (6.6) is

$$J \equiv \kappa^2 = \sum_{i \neq j=1}^{n} (s_{ij} - U_{ij})^2.$$

The descent vectors are determined as follows:

$$\begin{aligned} \frac{\partial J}{\partial u_{ar}} &= -2\{\sum_{i=1}^{n} (s_{ia} - U_{ia}) \sum_{k=1}^{K} w_{kr} u_{ik} \\ &+ \sum_{j=1}^{n} (s_{aj} - U_{aj}) \sum_{l=1}^{K} w_{rl} u_{jl}\}, \\ & a = 1, \cdots, n, \quad r = 1, \cdots, K. \end{aligned} \tag{6.7}$$

$$\frac{\partial J}{\partial w_{bc}} = -2 \sum_{i \neq j=1}^{n} (s_{ij} - U_{ij}) u_{ib} u_{jc}, \quad b, c = 1, \cdots, K. \tag{6.8}$$

The following method is used to find the solutions $\hat{U}$ and $\hat{W}$.

(Step 1) Fix K, $2 \leq K < n$.

(Step 2) Initialize $U(0) = (u_{ik}(0))$, $(i = 1, \cdots, n; \; k = 1, \cdots, K)$, where $u_{ik}(0)$ is generated by uniform pseudorandom numbers in the interval $[0, 1]$ and satisfy $\sum_{k=1}^{K} u_{ik}(0) = 1$. Initialize $W(0) = (w_{kl}(0))$, $(k, l = 1, \cdots, K)$, where $0 \leq w_{kl}(0) \leq 1$. Set the step number $q = 0$.

(Step 3) Set $q = q + 1$. Calculate the value of $\frac{\partial J}{\partial u_{ik}(q-1)}$ using equation (6.7) and $W(q-1)$ which is obtained in Step 2. Find the optimal solution with respect to the direction of the descent vector by using one dimensional direct search. That is update u_{ik} by using the expression,

$$u_{ik}(q) = u_{ik}(q-1) - \lambda \left(\frac{\partial J}{\partial u_{ik}(q-1)} \right),$$

where $\lambda > 0$ is the step size.

(Step 4) Calculate the value of $\frac{\partial J}{\partial w_{kl}(q-1)}$ by using equation (6.8) and $U(q)$ obtained in Step 3. Find the optimal solution with respect to the direction of the descent vector by using a one dimensional direct search. That is update w_{kl} by using the next expression,

$$w_{kl}(q) = w_{kl}(q-1) - \lambda \left(\frac{\partial J}{\partial w_{kl}(q-1)} \right),$$

where $\lambda > 0$ is the step size.

(Step 5) Calculate the value of $\|U(q) - U(q-1)\|$, where $\| \cdot \|$ shows the norm of matrix.

(Step 6) If $\|U(q) - U(q-1)\| < \varepsilon$, then stop, or otherwise go to Step 3.

6.1.3 Self-Organized Fuzzy Clustering Methods

When we obtain the result for the degree of belongingness for the fuzzy clustering model by using the above algorithm, there are times when we cannot obtain a suitable result as the solution is a local minimum solution. We present the new fuzzy clustering method using the result of the above fuzzy clustering.

The method consists of the following three steps:

(Step 1) Apply the similarity data for model (5.7). If the similarity data is symmetric, we can obtain the symmetric matrix for $W = (w_{kl})$. Model (5.7) is a generalized model of equation (6.3). Obtain the solutions $\hat{U} = (\hat{u}_{ik})$ and $\hat{W} = (\hat{w}_{kl})$.

(Step 2) Using the obtained $\hat{U}$, recalculate the following similarity:

$$\tilde{\hat{s}}_{ij} = \frac{1}{\sum_{k=1}^{K} (\hat{u}_{ik} - \hat{u}_{jk})^2} s_{ij}, \quad i,j = 1, \cdots, n. \tag{6.9}$$

Using $\tilde{\hat{s}}_{ij}$, go back to Step 1 and obtain the new result for $\tilde{\hat{U}}$ and $\tilde{\hat{W}}$.

(Step 3) Evaluate the fitness shown in equation (6.6) using $\tilde{\hat{U}}$ and $\tilde{\hat{W}}$ and compare with the fitness obtained by using $\hat{U}$ and $\hat{W}$. Repeat Steps 1 to 3.

Equation (6.9) shows that if objects i and j have a similar degree of belongingness for the clusters obtained. That is, if $\hat{u}_{ik}$ and $\hat{u}_{jk}$ are similar to each other, then the similarity between objects i and j becomes larger. This similarity is self organizing according to the degree of belongingness for each of the clusters obtained in each iteration.

Next we show another self-organized method using the FANNY method [27]. The following self-organized dissimilarity is proposed [54]:

$$\tilde{d}_{ij} = \sum_{k=1}^{K} (\hat{u}_{ik} - \hat{u}_{jk})^2 d_{ij}. \tag{6.10}$$

Where d_{ij} is the observed dissimilarity between i-th and j-th objects, and $\tilde{d}_{ij}$ shows the self-organized dissimilarity. $\hat{u}_{ik}$ is the estimated degree of belongingness of an object i to a cluster k which minimizes equation (1.4) when $m = 2$ and satisfying the condition (1.1). Using this condition, the following relationship is satisfied:

$$0 < \sum_{k=1}^{K} (\hat{u}_{ik} - \hat{u}_{jk})^2 \leq 2. \tag{6.11}$$

Equation (6.10) shows that if objects i and j have a similar degree of belongingness for the obtained clusters, that is, if $\hat{u}_{ik}$ and $\hat{u}_{jk}$ are similar to each other, then the dissimilarity $\tilde{d}_{ij}$ between objects i and j becomes smaller monotonously due to condition (6.11). This dissimilarity is self organizing according to the degree of belongingness for the clusters. The concept of self-organizing is that the objects organize adaptable dissimilarities among themselves, according to the similarity of their clustering situations. In each iteration of the fuzzy clustering algorithm, we use self-organized dissimilarity. The method consists of the following three steps:

(Step 1) Apply the dissimilarity data for equation (1.4) when $m = 2$. Obtain the solution $\hat{U} = (\hat{u}_{ik})$.

(Step 2) Using the obtained $\hat{U}$, recalculate the self-organized dissimilarity shown in equation (6.10). Using $\tilde{d}_{ij}$ in equation (6.10), return to Step 1 and obtain the new result of $\tilde{\hat{U}} = (\tilde{\hat{u}}_{ik})$.

(Step 3) If $\|\tilde{\hat{U}} - \hat{U}\| < \varepsilon$ then stop. Otherwise, repeat Steps 1 to 3.

6.2 Self-Organized Hybrid Method of Fuzzy Clustering and MDS

6.2.1 Multidimensional Scaling (MDS)

Multidimensional scaling is a method for capturing efficient information from observed dissimilarity data by representing the data structure in a lower dimensional spatial space. As a metric MDS (principal coordinate analysis), the following model [18], [31] has been proposed.

$$d_{ij} = \{\sum_{\lambda=1}^{R} d^{\gamma}(x_{i\lambda}, x_{j\lambda})\}^{\frac{1}{\gamma}} + \varepsilon_{ij}. \tag{6.12}$$

In equation (6.12), d_{ij} is an observed dissimilarity between objects i and j. $x_{i\lambda}$ is a point of an object i with respect to a dimension λ in a R dimensional configuration space. ε_{ij} is an error. $d^{\gamma}(x_{i\lambda}, x_{j\lambda})$ shows dissimilarity between objects i and j with respect to a dimension λ. Usually $d^{\gamma}(x_{i\lambda}, x_{j\lambda}) = | x_{i\lambda} - x_{j\lambda} |^{\gamma}$. That is, MDS finds a R dimensional scaling (coordinate) $(x_{i1}, \cdots, x_{iR})$ and throws light on the structure of the similarity relationship among the objects. This is done by representing the observed d_{ij} as the distance between a point $(x_{i\lambda})$ and a point $(x_{j\lambda})$ in a R dimensional space. In equation (6.12), we use the Euclidean distance where $\gamma = 2$. Since we use the Euclidean distance, the results of equation (6.12) is equivalent to R principal components in principal component analysis.

This is implemented using the assumption:

$$\sum_{i=1}^{n} x_{i\lambda} = 0, \ \ \forall \lambda, \tag{6.13}$$

due to the double centering procedure.

6.2.2 Hybrid Technique of Fuzzy Clustering and Multidimensional Scaling based on Self-Organized Dissimilarity

In metric MDS, under condition (6.13) and using the Young & Householder theorem [67], we can rewrite equation (6.10) as:

$$\begin{aligned} \tilde{d}_{ij} &= \sum_{k=1}^{K} (\hat{u}_{ik} - \hat{u}_{jk})^2 d_{ij} \\ &\simeq \sum_{k=1}^{K} (\hat{u}_{ik} - \hat{u}_{jk})^2 \sum_{\lambda=1}^{R} (\hat{\hat{x}}_{i\lambda} - \hat{\hat{x}}_{j\lambda})^2, \end{aligned} \tag{6.14}$$

where R eigen-values are sufficiently large over n eigen-values of XX', where $X = (x_{i\lambda})$. $\hat{\hat{x}}_{i\lambda}$ is an estimate of $x_{i\lambda}$.

As can be seen in equation (6.14), due to the spatial representation through the points $x_{i\lambda}$ in a R dimensional configuration space, we need to consider not only the monotonousness of the dissimilarity in equation (6.10), but also the interaction between a distance space consisting of $\hat{u}_{ik}$ and a distance space consisting of $\hat{\hat{x}}_{i\lambda}$.

Suppose

$$\hat{s}_{ij} \equiv \sum_{k=1}^{K} \hat{u}_{ik}\hat{u}_{jk}, \text{ and } \hat{\hat{s}}_{ij} \equiv \sum_{\lambda=1}^{R} \hat{\hat{x}}_{i\lambda}\hat{\hat{x}}_{j\lambda}.$$

Then equation (6.14) is shown as:

$$\tilde{d}_{ij} = (\hat{s}_{ii} + \hat{s}_{jj})(\hat{\hat{s}}_{ii} + \hat{\hat{s}}_{jj}) + 4\hat{s}_{ij}\hat{\hat{s}}_{ij} - 2\{(\hat{s}_{ii} + \hat{s}_{jj})\hat{\hat{s}}_{ij} + (\hat{\hat{s}}_{ii} + \hat{\hat{s}}_{jj})\hat{s}_{ij}\}. \tag{6.15}$$

We normalize $\hat{u}_{ik}$ and $\hat{\hat{x}}_{i\lambda}$ for each object as:

$$u_{ik}^* \equiv \frac{\hat{u}_{ik} - \frac{1}{K}}{\sigma_i^{(u)}}, \; x_{i\lambda}^* \equiv \frac{\hat{\hat{x}}_{i\lambda} - \bar{x}_i}{\sigma_i^{(x)}}, \tag{6.16}$$

$$\sigma_i^{(u)} = \sqrt{\frac{\sum_{k=1}^{K}(\hat{u}_{ik} - \frac{1}{K})^2}{K-1}}, \; \sigma_i^{(x)} = \sqrt{\frac{\sum_{\lambda=1}^{R}(\hat{\hat{x}}_{i\lambda} - \bar{x}_i)^2}{R-1}}, \; \bar{x}_i = \frac{\sum_{\lambda=1}^{R}\hat{\hat{x}}_{i\lambda}}{R}.$$

Then equation (6.15) is

$$\tilde{d}_{ij}^* = 4\{1 + s_{ij}^*\hat{s}_{ij}^* - (s_{ij}^* + \hat{s}_{ij}^*)\},$$

since

$$s_{ii}^* = \hat{s}_{ii}^* = 1, \; \forall i,$$

where

$$s_{ij}^* \equiv \sum_{k=1}^{K} u_{ik}^* u_{jk}^*, \; \hat{s}_{ij}^* \equiv \sum_{\lambda=1}^{R} x_{i\lambda}^* x_{j\lambda}^*.$$

We redefine the following self-organized dissimilarity as:

$$\hat{d}_{ij} \equiv \frac{\tilde{d}_{ij}^*}{4} - (1 + s_{ij}^*\hat{s}_{ij}^*) = -(s_{ij}^* + \hat{s}_{ij}^*). \tag{6.17}$$

In order to avoid $\hat{d}_{ij} < 0$, we use the following transformation:

$$\tilde{\hat{d}}_{ij} = \frac{1}{1 + \exp^{-\hat{d}_{ij}}}. \tag{6.18}$$

In equation (6.17), s^*_{ij} shows a correlation between u^*_{ik} and u^*_{jk} over K clusters. That is, if the classification structures of objects i and j are similar to each other in R^K, then s^*_{ij} becomes larger. Moreover, $\hat{s}^*_{ij}$ shows similarity between the points of objects i and j in R^R. That is, $\hat{d}_{ij}$ is simultaneously influenced by the similarity of the degree of belongingness and the similarity of points in the configuration space. Also, the monotonic relationship between the similarity and the dissimilarity is satisfied by equation (6.17). The method consists of the following steps:

(Step 1) Apply the dissimilarity data for equation (1.4) when $m = 2$. Obtain the solution $\hat{U} = (\hat{u}_{ik})$.

(Step 2) Apply the dissimilarity data for equation (6.12) when $\gamma = 2$ and obtain the estimate for the configuration $\hat{\hat{X}} = (\hat{\hat{x}}_{i\lambda})$.

(Step 3) Using $\hat{U} = (\hat{u}_{ik})$ and $\hat{\hat{X}} = (\hat{\hat{x}}_{i\lambda})$, normalize $\hat{u}_{ik}$ and $\hat{\hat{x}}_{i\lambda}$ for each object using equation (6.16) and obtain $U^* = (u^*_{ik})$ and $X^* = (x^*_{i\lambda})$ respectively.

(Step 4) Using the obtained U^* and X^*, calculate the self-organized dissimilarity shown in equation (6.18).

(Step 5) Using $\tilde{\hat{d}}_{ij}$ in equation (6.18), apply the MDS shown in equation (6.12) when $\gamma = 2$ and obtain the estimate for the configuration $\tilde{\tilde{X}} = (\tilde{\tilde{x}}_{i\lambda})$.

6.3 Numerical Examples for Self-Organized Fuzzy Clustering

We use the artificially created test data shown in figure 3.3. The results are shown in table 6.1. This table shows the results of the degree of belongingness of each object to clusters C_1 and C_2. The conventional method shown in equation (5.7) and the self-organized fuzzy clustering method using equation (6.9) are used. The number of clusters is 2. This table shows the similarity between the two results. Both of the results are adaptable considering the data feature shown in figure 3.3. That is, we can obtain the two groups for both of the results.

Table 6.1 Results using the Additive Fuzzy Clustering Model and the Self-Organized Fuzzy Clustering Method

Objects	Conventional Model		Proposed Method	
	C_1	C_2	C_1	C_2
o_1	0.91	0.09	1.00	0.00
o_2	0.76	0.24	1.00	0.00
o_3	0.97	0.03	1.00	0.00
o_4	0.86	0.14	1.00	0.00
o_5	0.02	0.98	0.00	1.00
o_6	0.29	0.71	0.00	1.00
o_7	0.00	1.00	0.00	1.00
o_8	0.17	0.83	0.00	1.00

Table 6.2 shows the results of the fitness shown using equation (6.6). From this, we see that a better result from the use of the self-organized fuzzy clustering method is obtained using equation (6.9).

Table 6.2 Comparison of Fitness

Fitness	κ^2	$\tilde{\kappa}^2$
	0.053	0.0001

In table 6.2, κ^2 and $\tilde{\kappa}^2$ are as follows:

$$\kappa^2 = \frac{\sum_{i \neq j=1}^{n} (s_{ij} - \sum_{k=1}^{K} \sum_{l=1}^{K} \hat{w}_{kl} \hat{u}_{ik} \hat{u}_{jl})^2}{\sum_{i \neq j=1}^{n} (s_{ij} - \bar{s}_{ij})^2}, \quad \bar{s}_{ij} = \frac{\sum_{i \neq j=1}^{n} s_{ij}}{n(n-1)}.$$

$$\tilde{\kappa}^2 = \frac{\sum_{i \neq j=1}^{n} (\tilde{\hat{s}}_{ij} - \sum_{k=1}^{K} \sum_{l=1}^{K} \tilde{\hat{w}}_{kl} \tilde{\hat{u}}_{ik} \tilde{\hat{u}}_{jl})^2}{\sum_{i \neq j=1}^{n} (\tilde{\hat{s}}_{ij} - \bar{\tilde{\hat{s}}}_{ij})^2}, \quad \bar{\tilde{\hat{s}}}_{ij} = \frac{\sum_{i \neq j=1}^{n} \tilde{\hat{s}}_{ij}}{n(n-1)}.$$

We now show an example which used real observations. Landsat data observed over the Kushiro marsh-land is used. The value of the data shows the amount of reflected light from the ground with respect to six kinds of light for 75 pixels. We get data from mountain area, river area, and city area. The 1st to the 25th pixels show mountain area, the 26th to the 50th are river area, and 51st to the 75th show the city area. The results are shown in figures 6.1 and 6.2. Figure 6.1 shows the result using the conventional fuzzy clustering model shown in equation (5.7) and figure 6.2 is the result of the self-organized fuzzy clustering method using equation (6.9). In these figures, the abscissa shows each pixel and the ordinate shows the degree of belongingness for each cluster. The number of clusters is 3. From these results, we see the result of the proposed method obtains a clearer result when compared with the result shown in figure 6.1.

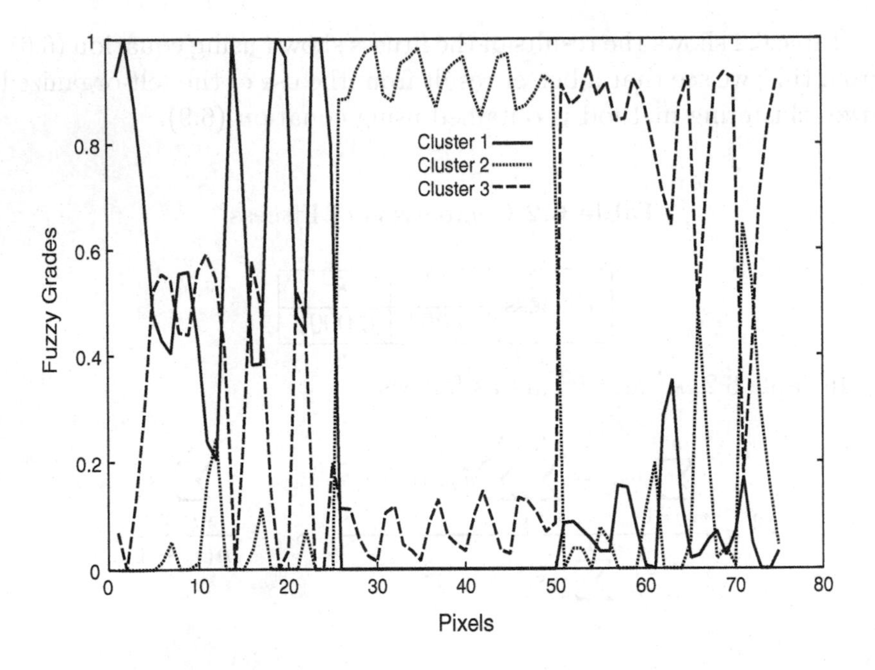

Fig. 6.1. Result of Landsat Data using the Fuzzy Clustering Model

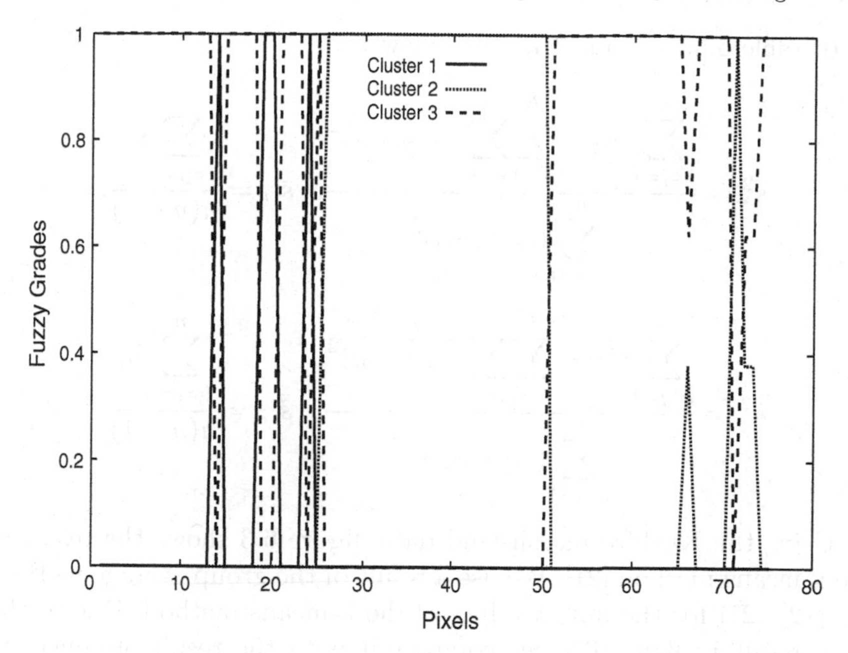

Fig. 6.2. Result of Landsat Data using the Self-Organized Fuzzy Clustering Method based on Equation (6.9)

Table 6.3 shows the comparison of the fitness for both the methods. From this table, we see that a better solution is obtained by using the self-organized fuzzy clustering method using equation (6.9).

Table 6.3 Comparison of the Fitness

Fitness	κ^2	$\tilde{\kappa}^2$
	0.14	0.07

In table 6.3, κ^2 and $\tilde{\kappa}^2$ are as follows:

$$\kappa^2 = \frac{\sum_{i \neq j=1}^{n} (s_{ij} - \sum_{k=1}^{K}\sum_{l=1}^{K} \hat{w}_{kl}\hat{u}_{ik}\hat{u}_{jl})^2}{\sum_{i \neq j=1}^{n} (s_{ij} - \bar{s}_{ij})^2}, \quad \bar{s}_{ij} = \frac{\sum_{i \neq j=1}^{n} s_{ij}}{n(n-1)}.$$

$$\tilde{\kappa}^2 = \frac{\sum_{i \neq j=1}^{n} (\tilde{\hat{s}}_{ij} - \sum_{k=1}^{K}\sum_{l=1}^{K} \tilde{\hat{w}}_{kl}\tilde{\hat{u}}_{ik}\tilde{\hat{u}}_{jl})^2}{\sum_{i \neq j=1}^{n} (\tilde{\hat{s}}_{ij} - \bar{\tilde{\hat{s}}}_{ij})^2}, \quad \bar{\tilde{\hat{s}}}_{ij} = \frac{\sum_{i \neq j=1}^{n} \tilde{\hat{s}}_{ij}}{n(n-1)}.$$

Using the Kushiro marsh-land data, figure 6.3 shows the result of the k-means method [21]. We use a result of the group average method [2], [12], [21] for the initial values of the k-means method. Due to the crisp result in figure 6.2, we compare it with the result obtained by the k-means method which is a hard clustering. As can be seen from the comparison between the results of figure 6.2 and figure 6.3, we obtain a different result by the use of the proposed method. If we obtain the same results, there is no reasons to use the self-organized method in order to obtain the defuzzified clustering result. We obtain the different features of the data as compared to the features obtained by the k-means method.

Figure 6.4 shows a comparison of the results of the proposed self-organized clustering method, the conventional fuzzy clustering model shown in equation (5.7), and the k-means method for each cluster. In these figures, the abscissa shows each pixel and the ordinate shows the degree of belongingness for each cluster. It can be seen that we obtain a more natural defuzzification where the degree of belongingness close to $\{0, 1\}$ arose from the fuzzy clustering result as compared to the result of the k-means method. The result is different from the result of the k-means method. Specifically, the difference in the result of the self-organized clustering from the k-means method is significant as shown in figure 6.4 (a). This figure shows the result of cluster C_1. The result of the self-organized clustering can maintain the features of the uncertainty classification structure obtained by the fuzzy clustering shown in model (5.7).

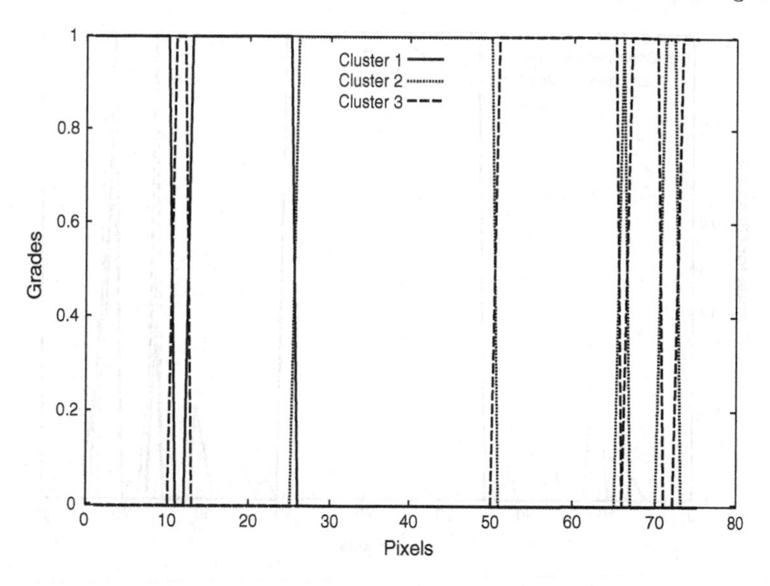

Fig. 6.3. Result of Landsat Data using the k-means Method

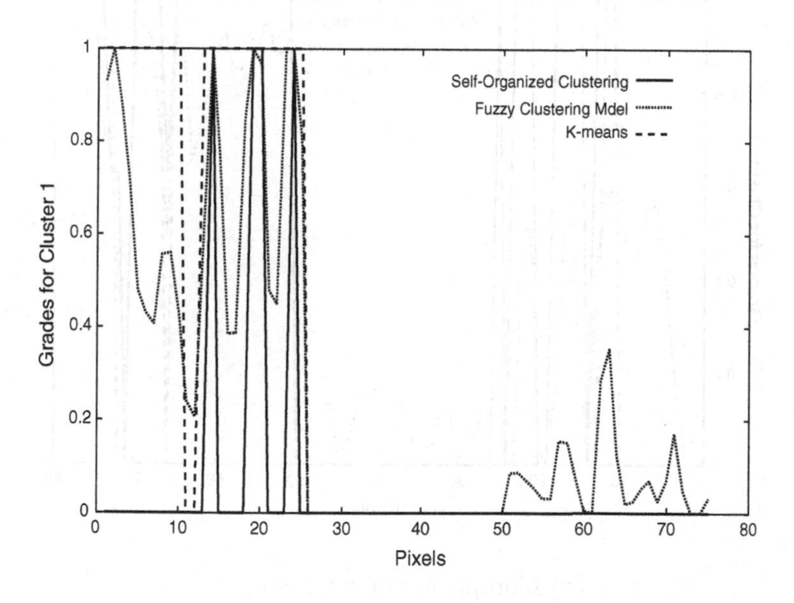

(a) Comparison for Cluster 1

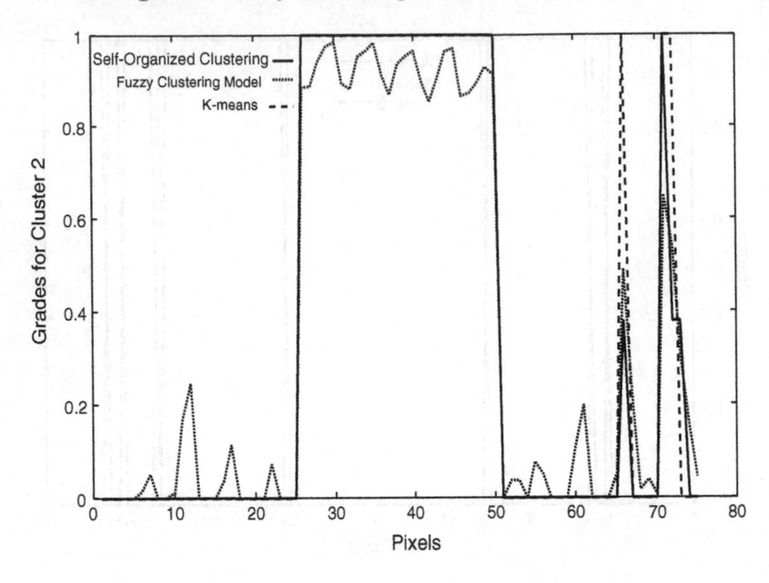

(b) Comparison for Cluster 2

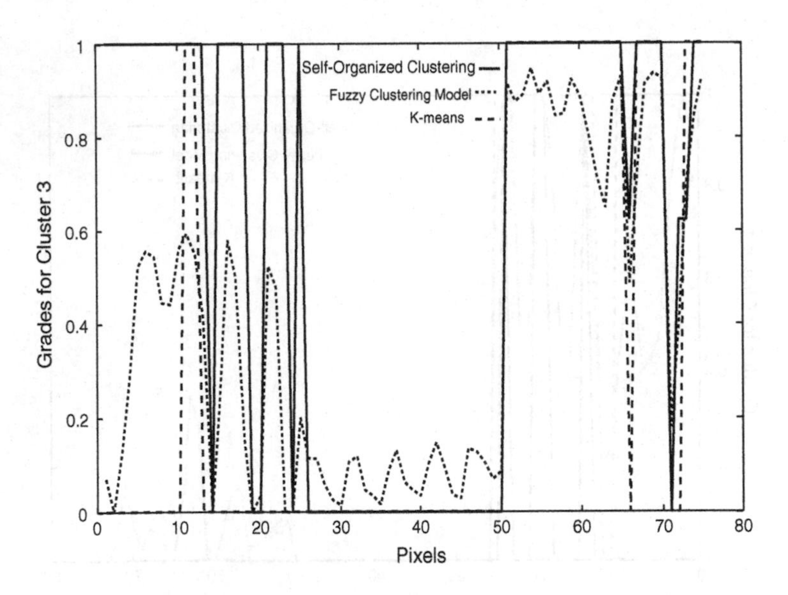

(c) Comparison for Cluster 3

Fig. 6.4. Comparison for Three Clusters over Three Methods

We now show the result of the hybrid technique using fuzzy clustering and multidimensional scaling. Figure 6.5 shows the result of MDS given by equation (6.12) using the artificially created test data shown in figure 3.3. Using data shown in figure 3.3, figure 6.6 is the result of the hybrid technique using fuzzy clustering and multidimensional scaling based on the self-organized dissimilarity shown in equation (6.18). In figures 6.5 and 6.6, the abscissa shows the values of the points with respect to the first dimension and the ordinate shows the values for the second dimension. These results are adaptable considering the data feature. That is, we can obtain two groups : a group of objects o_1, o_2, o_3 and o_4, and a group of objects o_5, o_6, o_7 and o_8.

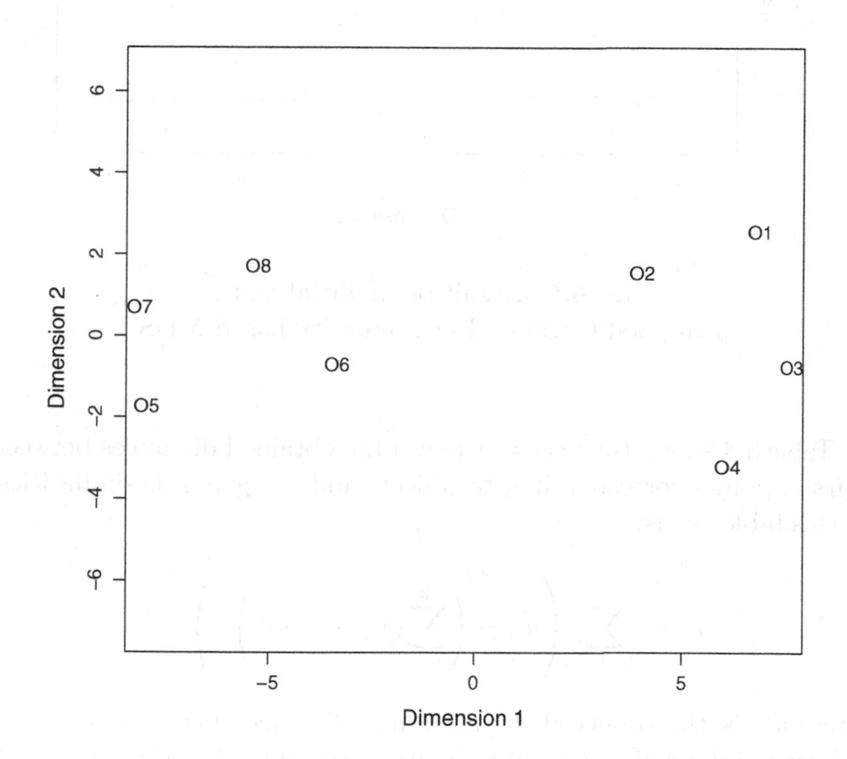

Fig. 6.5. Result for Artificial Data using MDS

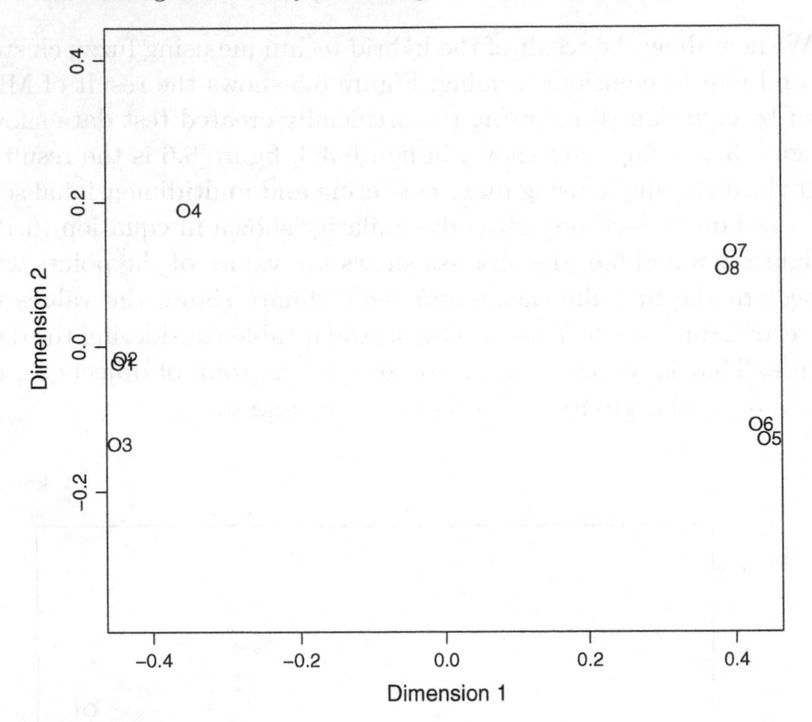

Fig. 6.6. Result of Artificial Data using Self-Organized Dissimilarity based MDS

Table 6.4 shows the fitness between the obtained distances between pairs of points corresponding to objects and the given dissimilarities. In this table, κ^2 is:

$$\kappa^2 \equiv \sum_{i \neq j = 1}^{n} \left(d_{ij} - \left(\sum_{\lambda=1}^{2} (\hat{\hat{x}}_{i\lambda} - \hat{\hat{x}}_{j\lambda})^2 \right)^{\frac{1}{2}} \right)^2 ,$$

where d_{ij} is the observed dissimilarity (distance) between a pair of objects i and j which is created by using the data shown in figure 3.3. $\hat{\hat{x}}_{i\lambda}$ shows the estimate of a point for an object i with respect to a dimension λ using the model shown in equation (6.12) when γ is 2. $\tilde{\kappa}^2$ shows:

$$\tilde{\kappa}^2 \equiv \sum_{i \neq j = 1}^{n} \left(\tilde{\hat{d}}_{ij} - \left(\sum_{\lambda=1}^{2} (\tilde{\hat{x}}_{i\lambda} - \tilde{\hat{x}}_{j\lambda})^2 \right)^{\frac{1}{2}} \right)^2 ,$$

where $\tilde{\tilde{x}}_{i\lambda}$ is the estimate of a point for an object i with respect to a dimension λ using the self-organized dissimilarity $\tilde{\hat{d}}_{ij}$ given by equation (6.18). From this table, we can see that a better result is obtained by using self-organized dissimilarity using the result of fuzzy clustering.

Table 6.4 Comparison of the Values of Fitness

Fitness	κ^2	$\tilde{\kappa}^2$
	0.78	0.58

We now use the Landsat data. Figure 6.7 shows the result of MDS shown in equation (6.12). Figure 6.8 is the result of the hybrid technique using fuzzy clustering and multidimensional scaling based on the self-organized dissimilarity given by equation (6.18).

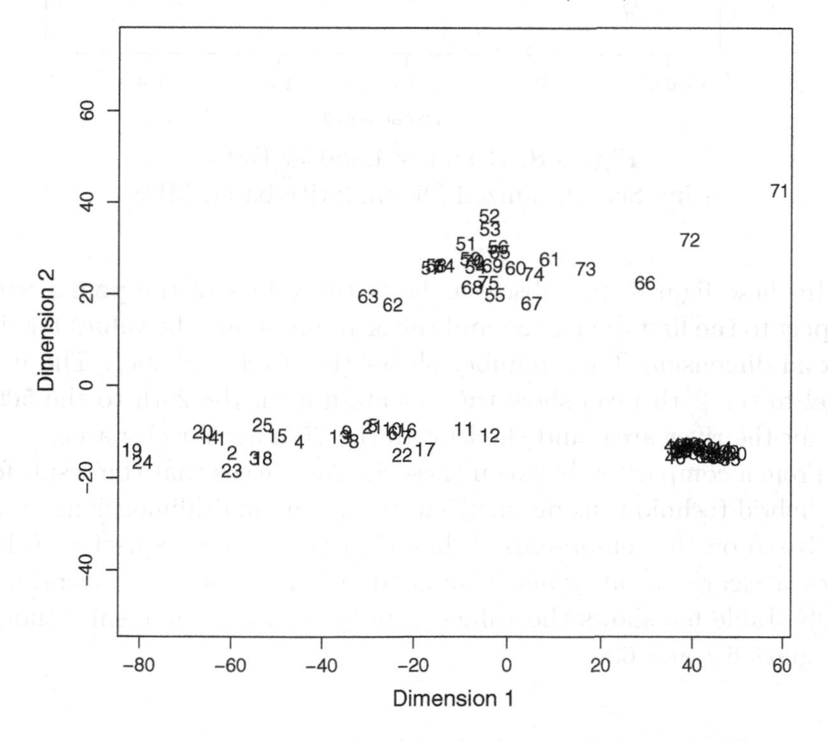

Fig. 6.7. Result of Landsat Data using MDS

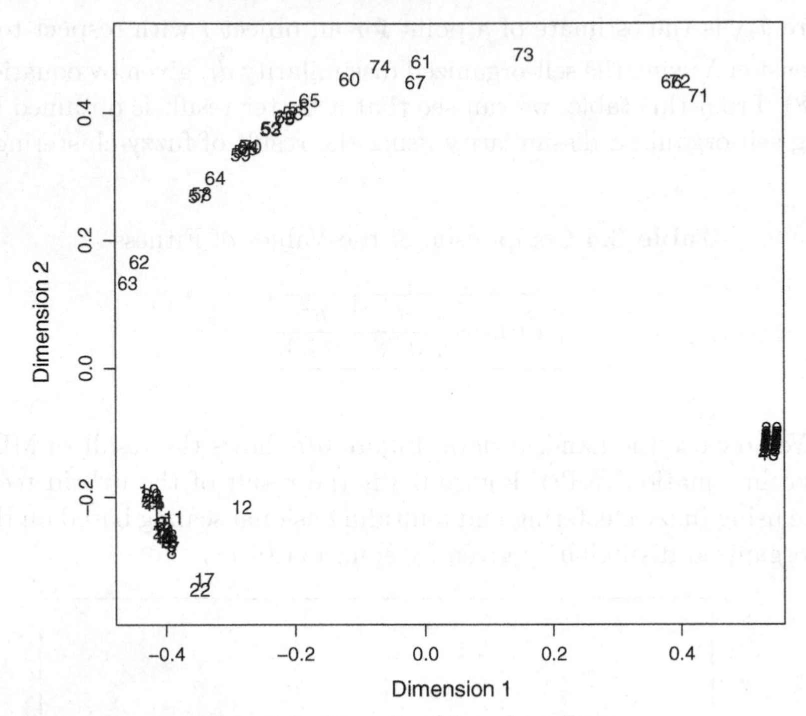

Fig. 6.8. Result of Landsat Data using Self-Organized Dissimilarity based MDS

In these figures, the abscissa shows the values of the points with respect to the first dimension and the ordinate shows the values for the second dimension. Each number shows the pixel's number. The first pixel to the 25th pixel show the mountain area, the 26th to the 50th are for the river area, and the 51st to the 75th are for city area.

From a comparison between these figures, we see that the result for the hybrid technique using fuzzy clustering and multidimensional scaling based on the self-organized dissimilarity shown in equation (6.18) gives a clearer result when compared to the result of conventional MDS. Table 6.5 shows the values of proportions of the results shown in figures 6.7 and 6.8.

Table 6.5 Comparison of the Values of Proportion

Proportion	Λ^*	Λ^{**}
	0.96	0.97

In this table the proportions are as follows:

$$\Lambda^* = \frac{\lambda_1^* + \lambda_2^*}{\sum_{l=1}^{6} \lambda_l^*},$$

where λ_l^*, $(l = 1, \cdots, 6)$ are eigen-values of the variance-covariance matrix of the estimates of points $\hat{\hat{X}} = (\hat{\hat{x}}_{i\lambda})$ in the six dimensional configuration space using the conventional MDS shown in equation (6.12) when $\gamma = 2$.

$$\Lambda^{**} = \frac{\lambda_1^{**} + \lambda_2^{**}}{\sum_{l=1}^{6} \lambda_l^{**}},$$

is the proportion of the result shown in figure 6.8 using the hybrid technique using fuzzy clustering and multidimensional scaling. This is based on the self-organized dissimilarity shown in equation (6.18). Where, λ_l^{**}, $(l = 1, \cdots, 6)$ are the eigen-values of the variance-covariance matrix of the estimates of points $\tilde{\tilde{X}} = (\tilde{\tilde{x}}_{i\lambda})$ in the six dimensional configuration space. This uses the hybrid technique with fuzzy clustering and multidimensional scaling. We see that the result of the hybrid technique with fuzzy clustering and multidimensional scaling based on the self-organized dissimilarity in equation (6.18) is better than the result of conventional MDS.

References

1. T. W. Anderson, *An Introduction to Multivariate Statistical Analysis*, Second ed., John Wiley & Sons, 1984.
2. M. R. Anderberg, *Cluster Analysis for Applications*, Academic Press, 1973.
3. Y. Baba and T. Nakamura, Time Dependent Principal Component Analysis, Measurement and Multivariate Analysis, S. Nishisato et al. eds., Springer, pp. 155-162, 2002.
4. J. C. Bezdek, *Pattern Recognition with Fuzzy Objective Function Algorithms*, Plenum Press, 1987.
5. J. C. Bezdek, J. Keller, R. Krisnapuram, and N. R. Pal, *Fuzzy Models and Algorithms for Pattern Recognition and Image Processing*, Kluwer Academic Publishers, 1999.
6. H. H. Bock and E. Diday eds., *Analysis of Symbolic Data*, Springer, 2000.
7. C. Brunsdon, S. Fotheringham, and M. Charlton, Geographically Weighted Regression-Modelling Spatial Non-Stationarity, *Journal of the Royal Statistical Society*, Vol. 47, Part 3, pp. 431-443, 1998.
8. N. Cristianini and J. Shawe-Taylor, *An Introduction to Support Vector Machines and Other Kernel-Based Learning Methods*, Cambridge University Press, 2000.
9. N. C. Da Cunha and E. Polak, Constrained Minimization under Vector Valued Criteria in Finite-Dimensional Space, *J. Math. Anal. and Appli.*, Vol. 19, No.1, pp. 103-124, 1967.
10. A. J. Dobson, *An Introduction to Generalized Linear Models*, Chapman and Hall, 1990.
11. N. R. Draper and H. Smith, *Applied Regression Analysis*, John Wiley & Sons, 1966.
12. B. Everitt, *Cluster Analysis*, Heinemann Educ. Books, 1974.
13. R. A. Fisher, The Use of Multiple Measurements in Taxonomic Problems, *Ann. Eugenics*, Vol. 7, No. 2, pp. 179-188, 1936.
14. H. Fujiwara, Methods for Cluster Analysis using Asymmetric Measures and Homogeneity Coefficient, *Beheviormetrika*, Vol. 7, No. 2, pp. 12-21, 1980 (in Japanese).
15. I. Gath and A. B. Geva, Unsupervised Optimal Fuzzy Clustering, *IEEE Trans. Patt. Anal. and Machine Intell.*, Vol. 11, pp. 773-781, 1989.
16. A. Gifi, *Nonlinear Multivariate Analysis*, John Wiley & Sons, 1990.
17. A. D. Gordon, A Review of Hierarchical Classification, *Journal of the Royal Statistical Society*, Series A, pp. 119-137, 1987.
18. J. C. Gower, Some Distance Properties of Latent Roots and Vector Methods used in Multivariate Analysis, *Biometrika*, Vol. 53, pp. 325-338, 1966.
19. T. J. Hastie and R. J. Tibshirani, *Generalized Additive Models*, Chapman & Hall, 1990.
20. R. Gunderson, Applications of Fuzzy ISODATA Algorithms to Star-Tracker Printing Systems, Proc. Triannual World IFAC Congress, pp. 1319-1323, 1978.

21. J. A. Hartigan, *Clustering Algorithms*, John Wiley and Sons, 1975.
22. R. J. Hathaway, J. W. Davenport, and J. C. Bezdek, Relational Duals of the C-Means Clustering Algorithms, *Pattern Recognition*, Vol. 22, pp. 205-212, 1989.
23. R. J. Hathaway and J. C. Bezdek, Switching Regression Models and Fuzzy Clustering, *IEEE Trans. Fuzzy Syst.*, Vo. 1, No. 3, pp. 195-204, 1993.
24. L. Hubert, Min and Max Hierarchical Clustering using Asymmetric Similarity Measures, *Psychometrika*, Vol. 38, pp. 63-72, 1973.
25. M. Ichino and H. Yaguchi, Generalized Minkowski Metrics for Mixed Feature-Type Data Analysis, *IEEE Transactions on Systems, Man, and Cybernetics*, Vol. 24, No. 4, pp. 698-708, 1994.
26. I. T. Jolliffe, *Principal Component Analysis*, Springer Verlag, 1986.
27. L. Kaufman and P. J. Rousseeuw, *Finding Groups in Data*, John Wiley & Sons, 1990.
28. H. A. L. Kiers, An Altering Least Squares Algorithm for Fitting the Two-and Three-Way DEDICOM Model and IDIOSCAL Model, *Psychometrika*, Vol. 54, pp. 515-521, 1989.
29. H. A. L. Kiers et al. eds., *Data Analysis, Classification, and Related Methods*, Springer, 2000.
30. T. Kohonen, Self-Organized Formation of Topologically Correct Feature Maps, *Biol. Cybern.*, Vol. 43, pp. 59-69, 1982.
31. J. B. Kruskal and M. Wish, *Multidimensional Scaling*, Sage Publications, 1978.
32. J. Mandel, *The Statistical Analysis of Experimental Data*, Dover Publications, 1964.
33. K. Menger, Statistical Metrics, *Proc. Nat. Acad. Sci.* USA, Vol. 28, pp. 535-537, 1942.
34. M. Sato and Y. Sato, Extended Fuzzy Clustering Models for Asymmetric Similarity, Fuzzy Logic and Soft Computing, B. Bouchon-Meunier et al. eds., World Scientific, pp. 228-237, 1995.
35. M. Sato and Y. Sato, On a General Fuzzy Additive Clustering Model, *International Journal of Intelligent Automation and Soft Computing*, Vol. 1, No. 4, pp. 439-448, 1995.
36. M. Sato, Y. Sato, and L. C. Jain, *Fuzzy Clustering Models and Applications*, Springer, 1997.
37. M. Sato-Ilic and Y. Sato, A Dynamic Additive Fuzzy Clustering Model, *Advances in Data Science and Classification, Springer-Verlag, A. Rizzi, M. Vichi, H. H. Bock, eds.*, pp. 117-124, 1998.
38. M. Sato-Ilic, On Dynamic Clustering Models for 3-way Data, *Journal of Advanced Computational Intelligence*, Vol. 3, No. 1, pp. 28-35, 1999.
39. M. Sato-Ilic and Y. Sato, Asymmetric Aggregation Operator and its Application to Fuzzy Clustering Model, *Computational Statistics & Data Analysis*, Vol. 32, pp. 379-394, 2000.
40. M. Sato-Ilic, On Evaluation of Clustering using Homogeneity Analysis, *IEEE International Conference on Systems, Man and Cybernetics*, pp. 3588-3593, 2000.
41. M. Sato-Ilic, Classification based on Relational Fuzzy C-Means for 3-Way Data, *International Conference on Enterprise Information Systems*, pp. 217-221, 2000.
42. M. Sato-Ilic, Homogeneity Measure for an Asymmetric Fuzzy Clustering Result, *The Third International Conference Intelligent Processing and Manufacturing of Materials*, 2001.
43. M. Sato-Ilic, On Clustering based on Homogeneity, *Joint 9th IFSA World Congress and 20th NAFIPS International Conference*, pp. 2505-2510, 2001.
44. M. Sato-Ilic, Fuzzy Regression Analysis using Fuzzy Clustering, *NAFIPS-FLINT 2002 International Conference*, pp. 57-62, 2002.

45. M. Sato-Ilic and T. Matsuoka, On an Application of Fuzzy Clustering for Weighted Regression Analysis, *The 4th ARS Conference of the IASC*, pp. 55-58, 2002.
46. M. Sato-Ilic, Fuzzy Cluster Loading for 3-Way Data, *Soft Computing and Industry -Recent Applications-*, Springer, pp. 349-360, 2002.
47. M. Sato-Ilic, On Kernel based Fuzzy Cluster Loadings with the Interpretation of the Fuzzy Clustering Result, *International Journal of Computational and Numerical Analysis and Applications*, Vol. 4, No. 3, pp. 265-278, 2003.
48. M. Sato-Ilic, Weighted Principal Component Analysis for Interval-Valued Data based on Fuzzy Clustering, *IEEE International Conference on Systems, Man and Cybernetics*, pp. 4476-4482, 2003.
49. M. Sato-Ilic, Asymmetric Kernel Fuzzy Cluster Loading, *Intelligent Engineering Systems through Artificial Neural Networks*, Vol. 13, pp. 411-416, 2003.
50. M. Sato-Ilic, On Fuzzy Clustering based Regression Models, *NAFIPS 2004 International Conference*, pp. 216-221, 2004.
51. M. Sato-Ilic, Weighted Principal Component Analysis based on Fuzzy Clustering, *Scientiae Mathematicae Japonicae*, Vol. 10, pp. 359-368, 2004.
52. M. Sato-Ilic, Fuzzy Clustering based Weighted Principal Component Analysis for Interval-Valued Data Considering Uniqueness of Clusters, *IEEE International Conference on Systems, Man and Cybernetics*, pp. 2297-2302, 2004.
53. M. Sato-Ilic, Self Organized Fuzzy Clustering, *Intelligent Engineering Systems through Artificial Neural Networks*, Vol. 14, pp. 579-584, 2004.
54. M. Sato-Ilic and T. Kuwata, On Fuzzy Clustering based Self-Organized Methods, *FUZZ-IEEE 2005*, pp. 973-978, 2005.
55. M. Sato-Ilic, On Features of Fuzzy Regression Models based on Analysis of Residuals, *Intelligent Engineering Systems through Artificial Neural Networks*, Vol. 15, pp. 341-350, 2005.
56. M. Sato-Ilic and J. Oshima, On Weighted Principal Component Analysis for Interval-Valued Data and Its Dynamic Feature, *International Journal of Innovative Computing, Information and Control*, Vol. 2, No. 1, pp. 69-82, 2006.
57. B. Scholkopf, A. Smola and K. Muller, Nonlinear Component Analysis as a Kernel Eigenvalue Problem, *Neural Computation*, Vol. 10, pp. 1299-1319, 1998.
58. B. Scholkopf and A. J. Smola, *Learning with Kernels*, The MIT Press, 2002.
59. B. Schweizer and A. Sklar, *Probabilistic Metric Spaces*, North-Holland, 1983.
60. Y. Takane and T. Shibayama, Principal Component Analysis with External Information on Both Subjects and Variables, *Psychometrika*, Vol. 56, pp. 97-120, 1991.
61. Y. Takane, Seiyakutuki Syuseibun Bunseki, Asakura Syoten, 1995 (in Japanese).
62. H. Tanaka and J. Watada, Possibilistic Linear Systems and their Application to the Linear Regression Model, *Fuzzy Sets and Systems*, Vol. 27, pp. 275-289, 1988.
63. M. M. Van Hulle, *Faithful Representations and Topographic Maps -From Distortion to Information based Self-Organization*, John Wiley & Sons, 2000.
64. W. N. Venables and B. D. Ripley, *Modern Applied Statistics with S-PLUS*, Springer, 1999.
65. J. Vesanto and E. Alhoniemi, Clustering of the Self-Organizing Map, *IEEE Transactions on Neural Networks*, Vol. 11, No. 3, pp. 586-600, 2000.
66. Y. Yamanishi and Y. Tanaka, Kansu Data No Chiriteki Omomituki Jyukaiki Bunseki (in Japanese), *Japanese Society of Computational Statistics*. pp. 1-4, 2001.
67. G. Young and A. S. Householder, Discussion of a Set of Points in Terms of their Mutual Distances, *Psychometrika*, Vol. 3, pp. 19-22, 1938.

68. L. A. Zadeh, Fuzzy Sets, *Inform. Control.*, Vol. 8, pp. 338-353, 1965.
69. Annual Report of Automated Meteorological Data Acquisition System (AMeDAS), Meteorological Business Support Center, 2000.
70. NTT-east, http://www.ntt-east.co.jp/info-st/network/traffic/, 1998.

Index